THE
MAGIC
OF
MATHEMATICS

Discovering the Spell of Mathematics

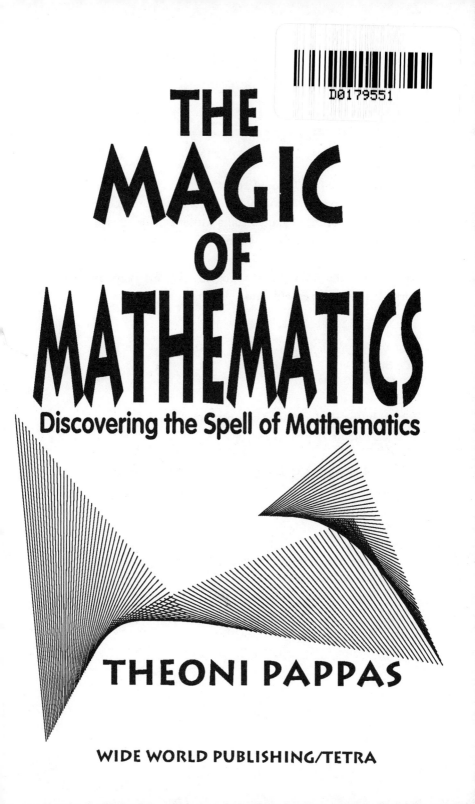

THEONI PAPPAS

WIDE WORLD PUBLISHING/TETRA

Portions of this book have appeared in previously published works, but were too essential to not be included.

Wide World Publishing/Tetra
P.O. Box 476
San Carlos, CA 94070

Printed in the United States of America.
Second Printing, October 1994.

Library of Congress Cataloging-in-Publication Data

Pappas, Theoni ,
The magic of mathematics : discovering the spell of mathematics / Theoni Pappas.
p. cm.
Includes bibliographical references and index ,
ISBN 0-933174-99-3
1. Mathematics- -Popular works. I. Title.
QA93 , P368 1994
510--dc20 94-11653
CIP

*This book is dedicated to
mathematicians
who have created and
are creating
the magic of mathematics.*

CONTENTS

PREFACE

You don't have to solve problems or be a mathematician to discover the magic of mathematics. This book is a collection of ideas — ideas with an underlying mathematical theme. It is not a textbook. Do not expect to become proficient in a topic or find an idea exhausted. *The Magic of Mathematics* delves into the world of ideas, explores the spell mathematics casts on our lives, and helps you discover mathematics where you least expect it.

Many think of mathematics as a rigid fixed curriculum. Nothing could be further from the truth. The human mind continually creates mathematical ideas and fascinating new worlds independent of our world — and *presto* these ideas connect to our world almost as if a magic wand had been waved. The way in which objects from one dimension can disappear into another, a new point can always be found between any two points, numbers operate, equations are solved, graphs produce pictures, infinity solves problems, formulas are generated — all seem to possess a magical quality.

Mathematical ideas are figments of the imagination. Its ideas exist in alien worlds and its objects are produced by sheer logic and creativity.   A perfect square or circle exists in a mathematical world, while our world has only representations of things mathematical.

The topics and concepts which are mentioned in each chapter are by no means confined to that section. On the contrary, examples can easily cross over the arbitrary boundaries of chapters. Even if it were possible, it would be undesirable to restrict a mathematical idea to a specific area. Each topic is essentially self-contained, and can be enjoyed independently. I hope this book will be a stepping stone into mathematical worlds.

Print Gallery by M.C. Escher.

MATHEMATICS IN EVERYDAY THINGS

There is no branch of mathematics, however abstract,
which may not someday be applied to phenomena
*of the real world. — **Nikolai Lobachevsky***

So many things with which we come into contact in our daily
routines have a mathematical basis or connection. These range
from taking a plane flight to the shape of a manhole. Often when
one least expects, one finds mathematics is involved. Here is a
random sampling of such cases.

THE MATHEMATICS OF FLYING

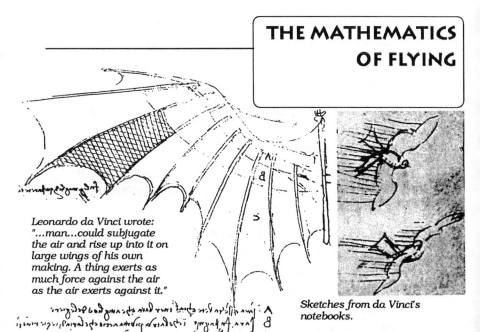

Leonardo da Vinci wrote:
"...man...could subjugate
the air and rise up into it on
large wings of his own
making. A thing exerts as
much force against the air
as the air exerts against it."

Sketches from da Vinci's
notebooks.

The grace and ease of the flight of birds have always tantalized human's desire to fly. Ancient stories from many cultures attest to interest in various flying creatures. Viewing hang gliders, one realizes that the flight of Daedalus and Icarus was probably not just a Greek myth. Today enormous sized aircrafts lift themselves and their cargo into the domain of the bird. The historical steps to achieve flight, as we now know it, has literally had its up and downs. Throughout the years, scientists, inventors, artists, mathematicians and other professions have been intrigued by the idea of flying and have developed designs, prototypes, and experiments in efforts to be airborne.

Here is a condensed outline of the history of flying:

•*Kites were invented by the Chinese (400-300 B.C.).*

•*Leonardo da Vinci scientifically studied the flight of birds and sketched various flying machines (1500).*

- *Italian mathematician Giovanni Borelli proved that human muscles were too weak to support flight (1680).*

- *Frenchmen Jean Pilâtre de Rozier and Marquis d' Arlandes made the first hot air balloon ascent (1783).*

- *British inventor, Sir George Cayley, designed the airfoil (cross-section) of a wing, built and flew (1804) the first model glider, and founded the science of aerodynamics.*

- *Germany's Otto Lilienthal devised a system to measure the lift produced by experimental wings and made the first successful manned glider flights between 1891-1896.*

- *In 1903 Orville and Wilbur Wright made the first engine powered propeller driven airplane flights. They experimented with wind tunnels and weighing systems to measure the lift and drag of designs. They perfected their flying techniques and machines to the point that by 1905 their flights had reached 38 minutes in length covering a distance of 20 miles!*

Here's how we get off the ground:
In order to fly, there are vertical and horizontal forces that must be balanced. Gravity (the downward vertical force) keeps us earthbound. To counteract the pull of gravity, lift (a vertical upward force) must be created. The shape of wings and the design of airplanes is essential in creating lift. The study of nature's design of wings and of birds in flight holds the key. It seems almost sacrilegious to quantify the elegance of the flight of birds, but without the mathematical and physical analyses of the components of flying, today's airplanes would never have left the ground. One does not always think of air as a substance, since it is invisible. Yet air is a medium, as water. The wing of an airplane, as well as the airplane itself, divides or slices the air as it passes through it. Swiss mathematician Daniel Bernoulli (1700-1782) discovered that as the speed of gas or fluid increases its pressure decreases. Bernoulli's principle[1] explains how the shape of a wing creates the lift force. The top of the wing is curved. This curve increases

the speed of air and thereby decreases the air pressure of the air passing over it. Since the bottom of the wing does not have this curve, the speed of the air passing under the wing is slower and thus its air pressure is higher. The high air pressure beneath the wing moves or pushes toward the low pressure above the wing, and thus lifts the plane into the air. The weight (the pull of gravity) is the vertical force that counteracts the lift of the plane.

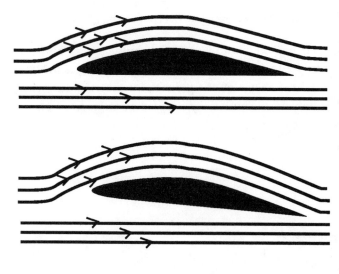

The wing's shape makes the distance over the top longer, which means air must travel over the top faster, making the pressure on the top of the wing lower than under the wing. The greater pressure below the wing pushes the wing up.

When the wing is at a steeper angle, the distance over the top is even longer, thereby increasing the lifting force.

Drag and thrust are the horizontal forces which enter the flying picture. Thrust pushes the plane forward while drag pushes it backwards. A bird creates thrust by flapping its wings, while a plane relies on its propellers or jets. For a plane to maintain a level and straight flight all the forces acting on it must equalize one another, i.e. be zero. The lift and gravity must be zero, while the thrust and drag must balance. During take off the thrust must be greater than the drag, but in flight they must be equal, otherwise the plane's speed would be continually increasing.

Viewing birds swooping and diving reveals two other flying factors.

When the speed of air over the top of the wing is increased, the lift will also increase.   By increasing the wing's angle to the approaching air, called the *angle of attack*, the speed over the top of the wing can be further increased. If this angle increases to approximately 15 or more degrees, the lift can stop abruptly and the bird or plane begins to fall instead of rising. When this takes place it is called the *angle of stall*. The angle of stall makes the air form vortexes on the top of the wing. These vibrate the wing causing the lift to weaken and the force of gravity to overpower the lift force.

Not having been endowed with the flying equipment of birds, humans have utilized mathematical and physical principles to lift themselves and other things off the ground. Engineering designs and features[2] have been continually adapted to improve an aircraft's performance.

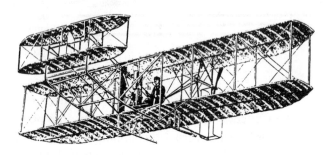

[1]Laws governing the flow of air for airplanes apply to many other aspects in our lives, such as skyscrapers, suspension bridges, certain computer disk drives, water and gas pumps, and turbines.

[2]The flaps and slots are changes adapted to the wing which enhance lift. The flap is a hinged section that when engaged changes the curvature of the wing and adds to the lift force. Slots are openings in the wing that delay the stall for a few degrees.

THE MATHEMATICS OF A TELEPHONE CALL

Every time you pick-up the telephone receiver to place a call, send a fax, or modem information — you are entering a phenomenally complicated and enormous network. The communication net that encompasses the globe is amazing. It is difficult to imagine how many calls are fielded and directed each day over this network. How does a system which is "broken-up" by varied systems of different countries and bodies of water operate? How does a single phone call find its way to someone in your city, state or another country?

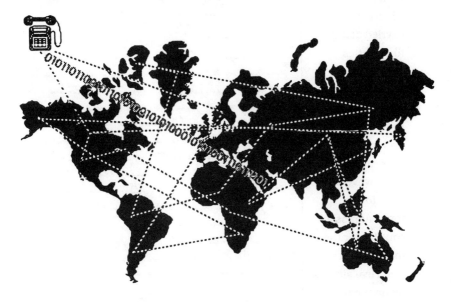

In the early years of the telephone, one picked-up the receiver and cranked the phone to get an operator. A local operator came on the line from the local switch board and said "number please", and from there connected you with the party you were trying to reach. Today the process has mushroomed as have the various methods used to convert and direct calls. Mathematics involving sophisticated types of linear programming, coupled with binary systems and codes, make sense out of a potentially precarious situation.

How does your voice travel? Your voice produces sounds which are converted in the receiver to electrical signals. Today these electrical impulses can be carried and converted in a variety of ways. They may be changed to laser light signals which are then carried along fiber optics cables[1], they may be converted to radio signals and transmitted over radio or microwave links from tower to tower across a country, or they may remain as electrical signals along the phone lines. Most of the calls connected in the USA are done by an automatic

Pages from the notebook of Alexander Graham Bell in which he writes of his first telephone message on his invention talking to his assistant, Mr. Watson. "I then shouted into M the following sentence: 'Mr. Watson — Come here— I want to see you.' To my delight he came and declared that he had heard and understood what I said."

switching system. Presently the electronic switching system is the fastest. Its system has a program which contains the needed information for all aspects of telephone operations while keeping track of which telephones are being used and which paths are available. Calls can be transmitted by electric currents at different frequencies or converted to digital signals. Either method enables multiple conversations to be transmitted along the same wires. The most modern systems convert calls into digital signals which are then encoded with a binary number sequence. The individual calls can thus travel "simultaneously" along the lines in a specified order until they are decoded for their destinations.

When a call is placed, the system chooses the best path for the call and sends a chain of commands to complete the circuitry. The entire process takes a fraction of a second. Ideally it would take a direct route to the other party— that would be desirable from the view point of the economics of distance and time. But if the direct line is at capacity servicing other calls, the new call must be sent along the best of the alternative routes. Here is where linear programming[2] comes into the picture. Visualize the telephone routing problem as a complex geometric solid with millions of facets. Each vertex represents a possible solution. The challenge is to find the best solution without having to calculate every one. In 1947, mathematician George B.

Danzig developed *the simplex method* to find the solution to complex linear programming problems. The *simplex method*, in essence, runs along the edges of the solid, checking one corner after another, while always heading for the best solution. As long as the number of possibilities is no more than 15,000 to 20,000, this method manages to find the solution efficiently. In 1984, mathematician Narendra Karmarkar discovered a method that drastically cuts down the time needed to solve very cumbersome linear programming problems, such as the best routes for telephone calls over long distances. The *Karmarkar algorithm* takes a short-cut by going through the middle of the solid. After selecting an arbitrary interior point, the algorithm warps the entire structure so that it reshapes the problem which brings the chosen point exactly into the center. The next step is to find a new point in the direction of the best solution and to warp the structure again, and bring the new point into the center. Unless the warping is done, the direction that appears to give the best improvement each time is an illusion.   These repeated transformations are based on concepts of projective geometry and lead rapidly to the best solution.

Today, the old telephone salutation "number please" takes on a double meaning. The once simple process of picking up your telephone receiver and placing a call, now sets into motion a vast and complicated network that relies on mathematics.

[1]Depending on the type of lines used, the number of "simultaneous" conversations can range from 96 to over 13000. Fiber optic systems can carry even more information than the traditional copper/aluminum cables.

[2]Linear programming techniques are used to solve a variety of problems. Usually the problems entail many conditions and variables. A simple case may be an agricultural problem: *A farmer wants to decide how to most effectively use his/her land to maximize production and profit.* Conditions and variables would involve such things as considering different crops, how much land each crop requires, how much yield each produces per acre, and how much revenue each brings when sold. To solve such a problem, one writes linear inequalities and/or equations for each condition and looks at a 2-dimensional graph of a polygonal region for the solution.

PARABOLIC REFLECTORS & YOUR HEADLIGHTS

When you flick the switch of your headlights from bright to dim, mathematics is at work. To be specific, the principles of a parabola do the trick. The reflectors behind the headlights are parabolic in shape. In fact, they are paraboloids (3-dimensional parabolas formed by rotating a parabola[1] about its axis of symmetry). The bright beam is created by a light source located at the focal point of the parabolic reflectors. Thus, the light rays travel out parallel to the parabola's axis of symmetry. When the lights are dimmed, the light source changes location. It is no longer at the focus, and as a result the light rays do not travel parallel to the axis. The low beams now point down and up. Those pointing up are shielded, so that only the downward low beams are reflected a shorter distance than the high beams.

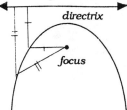

The parabola is an ancient curve that was discovered by Menaechmus (circa 375-325 B.C.) while he was trying to duplicate the cube. Over the centuries, new uses and discoveries involving the parabola have been made. For example, it was Galileo (1564-1642) who showed that a trajectile's path was parabolic. Today one can go into a hardware store and find a highly energy efficient parabolic electric heater which uses only 1000 watts but produces the same number of BTU thermal units as a heater that operates on 1500 watts.

---

[1]Parabola is the set of all points in a plane which are equidistant from a fixed point called its focus and a fixed line called its directrix.

COMPLEXITY & THE PRESENT

"The hours from seven 'til nearly midnight are normally quiet ones on the bridge. ... Beginning almost exactly at seven o'clock, ...it just looked as if everybody in Manhattan who owned a motorcar had decided to drive out on Long Island that evening."

As this excerpt from *The Law* by Robert M. Coates illustrates, sometimes things just seem to take place with no apparent reason. Nor is there a warning that a particular event is about to take place. We have all experienced such events and usually attributed them to "coincidence", since there were no apparent indicators to predict otherwise.

Complexity is an emerging science which may hold answers or at least explanations to such questions as:

How is it that

- the universe emerged out of the void ?
- cells know which organs and parts to become and when?
- on January 17, 1994 Los Angeles suffered an earthquake of unexpected magnitude and destruction?
- the Soviet Union's long reign over its satellite countries collapsed in such a short time?
- Yugoslavia was thrown suddenly into severe internal wars?
- a species that has not changed for millions of years suddenly experiences a mutation?
- for no apparent reason the stock market surges upward or plunges downward?

The list is endless. The underlying common factor of these events is that each represents a very complex system. A system governed by an enormous number and diversity of factors, which are delicately balanced, tittering between stability and chaos. The factors which act on such a system are ever growing and changing. Consequently, a complex system is always in a state of potential chaos i.e. at *the edge of chaos*. There seems to be a continual tug of war between order and chaos. *Spontaneous self-organizing dynamics* are an essential part of a complex system. It is the means by which the system regains equilibrium by changing and adapting itself to constantly changing factors/circumstances. Those studying this new science draw on a host of mathematical and scientific ideas, such as chaos theory, fractals, probability, artificial intelligence, fuzzy logic, etc. These scientists and mathematicians feel that today's mathematics, along with other tools and high tech innovations, are capable of creating a complexity framework that can impact major aspects of our global world, especially economics, the environment, and politics.

MATHEMATICS & THE CAMERA

Ever wonder about the f-stop number of a camera? Where did it get its name? How is it determined?   "f" stands for the mathematical term *factor*.   The brightness of the photographic image on film depends on the aperture and focal length of the lens. Photographers use what is known as the f-number system to relate focal length and aperture. The f-stop is calculated by measuring the diameter of the aperture and dividing it into the focal length of the lens. For example,

f4= 80mm lens/20mm aperture.

f1 6=80mm lens/5mm aperture.

We see the lens opening is smaller (the aperture decreases) as the f-stop number increases.   Working with f-stop numbers and shutter speeds, you can manually decide how much of the photograph you want in focus.

RECYCLING THE NUMBERS

Here mathematical units and symbols were used to get the point across about recycling paper!

- A ton of virgin paper ≠ a ton of recycled paper
- A ton of recycled paper uses 4102 kwh less energy.
- A ton of recycled paper uses 7000 gallons less water to produce.
- A ton of recycled paper produces 60 pounds less air pollution.
- A ton of recycled paper produces 3 cubic yards less solid waste.
- A ton of recycled paper uses less tax money for landfill.
- A ton of recycled paper uses 17 fewer logged trees.

— the numbers behind recycling and landfill —

- 37% of all landfill is comprised of paper.
- Only 29% of all newspapers produced are recycled by the consumer.
- 165 million cubic yards of landfill are needed for our paper wastes per year.
- 97% of the virgin forests of the continental USA have been cut down in the past 200 years.

I Was Once A Tree ...Newsletter, Spring 1990, Alonzo Printing, Hayward CA.

BICYCLES, POOL TABLES & ELLIPSES

The ellipse, along with other conic section curves, was studied by the Greeks as early as the 3rd century

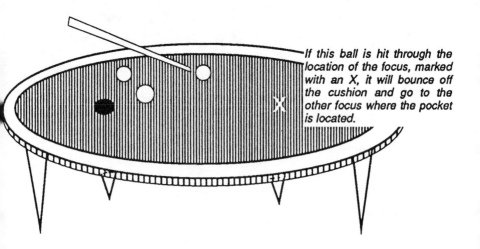

If this ball is hit through the location of the focus, marked with an X, it will bounce off the cushion and go to the other focus where the pocket is located.

B.C.. Most of us associate the ellipse with an angled circle or the orbital path of a planet, but elliptical shapes and properties also lend themselves to

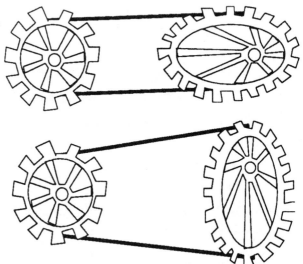

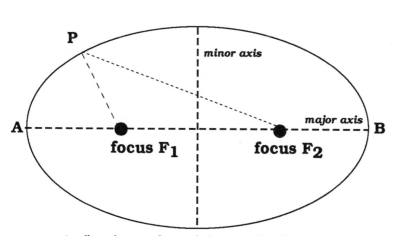

An ellipse has two foci, and the sum of the distances
from the foci to any point of the ellipse always equals
the length of its major axis, i.e. $|PF_1|+|PF_2|.=|AB|$.

contemporary nonscientific applications. Who would have
imagined that an ellipse would find itself in the design of bicycle
gears and pool tables? Today some bicycles have been
manufactured with a front elliptical gear and circular rear gears.
The drawing, on the previous page, illustrates how this design can
utilize the downward thrust of leg power and a quick upward return. *Elliptipools*, elliptical shaped pool tables, are designed to
utilize the reflection property of the ellipse's two foci. As illustrated
on the previous page, the elliptipool has one pocket located at one
of the two focus points of the ellipse. A ball hit so that it passes
through the ellipse's non-pocket focus will bounce off the side of
the table and travel the reflected path over to the pocket (the other
focus).

LOOKOUT FOR TESSELLATIONS

This Escher-like transformation by Mark Simonson illustrates the use of tessellations as a form of visual communication. This graphic appeared in *The Utne Reader* and on the cover of *Transactions,* a Metropolitan Transportation Communication publication. Reprinted courtesy of Mark Simonson. Bluesky Graphics, Minneapolis, MN.

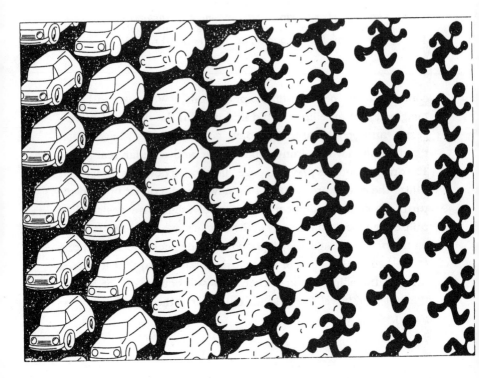

STAMPING OUT MATHEMATICS

One usually doesn't expect to encounter mathematical ideas on a trip to the post office, but here are a few of the stamps that have been printed with mathematical themes. These and many other ideas have appeared on such popular items as posters, television, T-shirts, post-its, mugs, bumper stickers, and stickers.

The Pythagorean theorem —Greece

LAS 10 FORMULAS MATEMATICAS QUE CAMBIARON LA FAZ DE LA TIERRA

The Pythagorean theorem —Nicaragua

Euler—Switzerland

Pascal—France

Möbius strip—Brazil

Bolyai—Rumania

Gauss—Germany

Mathematical Formulas —Israel

THE MOUSE'S TALE

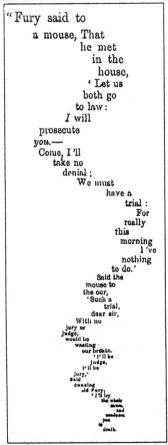

"Fury said to
a mouse, That
he met
in the
house,
'Let us
both go
to law:
I will
prosecute
you. —
Come, I'll
take no
denial ;
We must
have a
trial :
For
really
this
morning
I've
nothing
to do.'
Said the
mouse to
the cur,
'Such a
trial,
dear sir,
With no
jury or
judge,
would be
wasting
our breath.
'I'll be
judge,
I'll be
jury,'
Said
cunning
old Fury
'I'll try
the whole
cause,
and
condemn
you
to
death.

'Contrariwise,' continued Twee-
dledee, 'if it was so, it might be;
and if it were so, it would be: but
as it isn't, it ain't. That's logic.'

—Lewis Carrol
Alice in Wonderland

Charles Lutwidge Dodgson (1832-
1898) was a mathematician, lecturer
of mathematics, creator of puzzles
and games, renowned Victorian children's photographer and author of
mathematics books[1], children's stories, poetry and essays on social is-
sues. When writing children's tales
his pseudonym was Lewis Carroll. It
seems that Dodgson did not want to
be associated or connected with his
Lewis Carroll identity. In fact, he
would return mail addressed to Lewis
Carroll. When the Bodleian Library
cross-referenced Dodgson and Carroll, Dodgson took exception to the
connection.

Dodgson had a passion for word
games. In 1991, two teenagers from
New Jersey discovered a four-way pun
(both visual and verbal in nature) in the poem, *A Caucus-Race and*
a Long Tale from Chapter Two in *Alice in Wonderland*. Alice is told
the tale by a mouse in a poem shaped like a long tail. In addition
to this tale and tail link, students Gary Graham and Jeffrey Maid-
en discovered that when the poem was written in stanza form the
outline of a mouse was formed—each stanza had two short lines

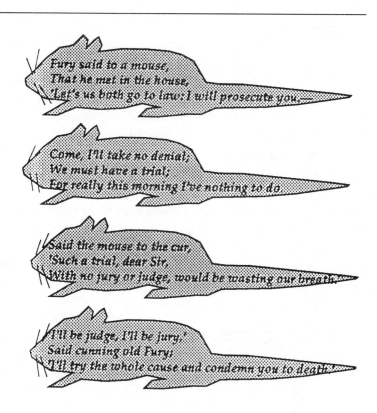

(the mouse's body) and a long third line (the mouse's tail). Lastly, they found that a *tail-rhyme* is a poetic structure defined by a pair of rhyming lines followed by another line of different length. Do you think Lewis Carroll planned all this intentionally?

---

[1] *Euclid and His Modern Rivals, An Elementary Treatise on Determinants, Alice in Wonderland, The Hunting of the Snark, Phantasmagoria and Other Poems, Through the Looking Glass* are of a few of Dodgson works.

A MATHEMATICAL VISIT

Not quite sure what to expect, I rang the doorbell. A voice asked me to please push the first five terms of the Fibonacci sequence.

Fortunately, I had done some research after my magazine assigned me the story on the home of the renowned mathematician, Selath.

I pushed 1, 1, 2, 3, 5 and the door slowly opened. As I passed through the doorway, I was struck by the catenary stone shaped archway independently suspended at the entrance. After a minute, Selath entered saying, "May I offer you something after your long drive?"

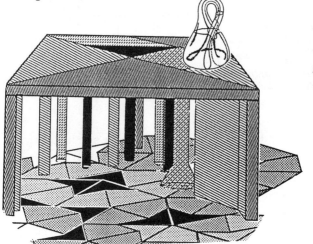

"I'd really appreciate a glass of cold water," I replied.

"Please come with me," he said, leading the way.

As I followed, I couldn't help noticing the many unique and unusual objects. In the kitchen, we came to a peculiar table with many legs. Selath pulled an equally unusual bottle from the refrigerator. I must have had a quizzical expression, for Selath began, "While you drink your water, we might as well start the tour here in the kitchen." As you noticed this table and bottle are not your everyday accessories. I use tangram tables for dining because its seven components can be rearranged into as many shapes as the tangram puzzle. Here in the kitchen it's made into a square shape today, while I've arranged the one in the living room into a triangle, since I am expecting two guests for dinner. The water container is what's known as a Klein

bottle— its inside and outside are one. If you look at the floor you'll notice only two shapes of tiles are used."

"Yes," I replied, " but the design doesn't seem to repeat anywhere."

"Very perceptive." Selath seemed pleased with my response. "These are Penrose tiles. These two shapes can cover a plane in a non-repeating fashion."

"Please continue," I urged. " I'm most anxious to see all the mathematical parts of your home."

"Well, actually almost all of my house is mathematical. Anywhere you see wall paper I've designed special tessellation patterns for walls á la Escher. Let's proceed to the Op room. Every item in here is an optical illusion. In fact, reality in this room is an illusion. Furniture, fixtures, photos, everything! For example, the couch is made from modulo cubes in black and white fabric stacked to give the feeling of an oscillating illusion. The sculpture in the middle was designed to show convergence and divergence, while the two stacked figures are exactly the same size.

This lamp's base, viewed from this location, makes the impossible tribar."

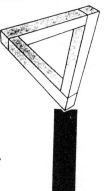

"Fascinating! I could spend hours discovering things in this room," I replied enthusiastically.

"Since we are on a tight schedule, let's move to the next room," Selath said as he led the way.

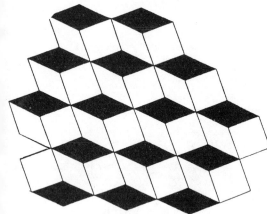

We entered a darkened room . "Watch your step here. Come this way to the parabolic screen," Selath directed.

As I peered into the disc a moving scene appeared. "Is this a video camera?" I asked.

"Oh, no," Selath laughed, "I call it my antique surveillance system. The lens above the hole captures light in the day time and rotates to project scenes outside my home, much the same way a camera would. It is called a camera obscura. I have a special lens for night viewing."

I was busily taking notes, realizing I would have much additional research to do before writing my article.

Glancing around I noted, "Your fluorescent clock seems to be off." My watch read 5:30 pm while his read 21:30.

"No, it's just that I have the 24 hours of the day arranged in base

eight because I'm working on eight hour cycles this week. So 24:00 hours would be 30:00 hours, 8:00 would be 10:00, and so on." Selath explained.

"Whatever works best for you," I replied, a bit confused.

"Now, let's go to the master bedroom."

And off we went, passing all sorts of shapes and objects I'd never seen in a home before.

"The master bedroom has a semi-spherical skylight in addition to movable geodesic skylights. They are designed to optimize the use of solar energy."

"Marvelous, but where is the bed?" I asked.

"Just push the button on this wooden cube, and you will see a bed unfold with a head board and two end tables."

"What a great way to make a bed, " I replied.

"There are many more things to see, but time is short. Let's go in the bathroom so you can see the mirrors over the basin. Come this way. Now lean forward."

To my surprise I saw an infinite number of images of myself repeated. The mirrors were reflecting back and forth into one another ad infinitum.

"Now turn around and notice this mirror. What's different about it?" Selath asked.

"My part is on the wrong side," I replied.

"To the contrary this mirror[1] lets you see yourself as you are really seen by others," Selath explained.

Just then the doorbell rang. The dinner guests had arrived. "Why don't you stay to dinner?" Selath asked. "You haven't seen the living room yet, and I'm sure you'll enjoy meeting my guests."

It was hard to conceal by enthusiasm. "But your table is set for three," I blurted.

"No problem. With the tangram table I can just rearrange a few parts and we'll have a rectangle.

---

[1]Made from two mirrors placed at right angles to each other. The right-angled mirrors are then positioned so that they will reflect your reflection.

THE EQUATION OF TIME

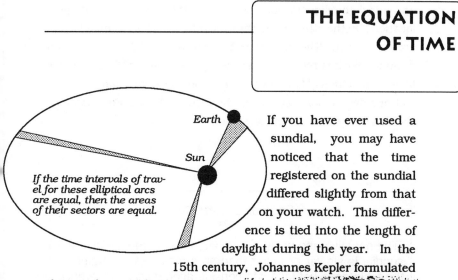

Earth

Sun

If the time intervals of travel for these elliptical arcs are equal, then the areas of their sectors are equal.

If you have ever used a sundial, you may have noticed that the time registered on the sundial differed slightly from that on your watch. This difference is tied into the length of daylight during the year. In the 15th century, Johannes Kepler formulated three laws that governed planetary motion. Kepler described how the Earth travels around the Sun in an elliptical orbit, and also explained that the line segment joining the Sun and the Earth sweeps out equal areas (sectors) in equal intervals of time along its orbit. The Sun is located at one of the foci of the ellipse thereby making each sector's area equal for a fixed time interval and the arc lengths of the sectors unequal. Thus the Earth's orbit speed varies along its path. This accounts for the variations in the lengths of daylight during different times of year. Sundials rely on daylight,

A 10th century pocket sundial. There are six months listed on each side. A stick is placed in the hole of the column with the current month.

and daylight depends on the time of the year and geographic
location. On the other hand, the time intervals of our other clocks
are consistent. The difference between a sundial's time and an
ordinary clock is referred to as *the equation of time*. The almanac
lists *the equation of time chart*, which indicates how many minutes
fast or slow the sundial is from the regular clocks. For example,
the chart may look like the one below.

Equation of Time Chart

(The negative and positive numbers indicate the minutes the
sundial is slower or faster than an ordinary clock. Naturally
the table does not take into consideration daylight difference
within time zones.)

DATE		VARIATION
Jan	1	- 3
	15	- 9
Feb	1	-13
	15	-14
Mar	1	- 3
	15	- 9
April	1	- 4
	15	0
May	1	+ 3
	15	+ 4
...		...

*If the sundial shows 11:50 on
May 15, its time should be
increased by 4 minutes to 11:54.*

WHY ARE MANHOLES ROUND?

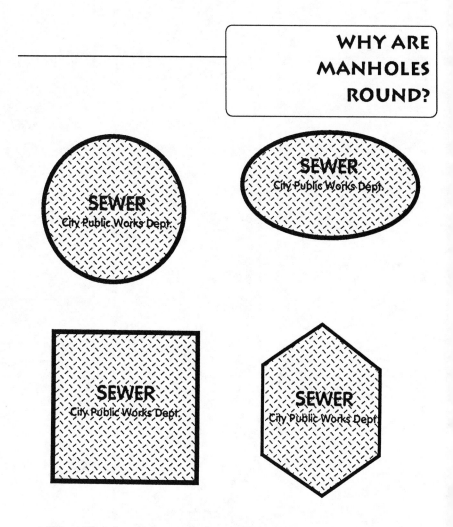

Why is the shape of a manhole circular?

Why not a square, rectangular, hexagonal, or elliptical shape?

Is it because a circle's shape is more pleasing?

There is a mathematical reason.

What is your explanation?

Metamorphosis III by M.C. Escher.

MAGICAL
MATHEMATICAL
WORLDS

How can it be that mathematics, a product of human thought independent of experience, is so admirably adapted to the objects of reality.
—*Albert Einstein*

Mathematics is linked and used by so many things in our world, yet delves in its own worlds— worlds so strange, so perfect, so totally alien to things of our world. A complete mathematical world can exist on the pin point of a needle or in the infinite set of numbers. One finds such worlds composed of points, equations, curves, knots, fractals, and so on. Until one understands how mathematical worlds and systems are formed, some of its worlds may seem contradictory. For example, one might ask how an infinite world can exist only on a tiny line segment, or a world be created using only three points.This chapter seeks to explore the magic of some of these mathematical worlds and delve into their domains.

$$1,2,3,4,5,6,7,8,9,10,11,12,13,...$$

As discussed later, the counting numbers form a mathematical world in themselves.

*Why, sometimes I've believed
as many as six impossible things
before breakfast.*
—Lewis Carroll

Little did Euclid know in 300 B.C. when he began to organize geometric ideas into a mathematical system that he was developing the first mathematical world. Mathematical worlds and their elements abound — here we find the world of arithmetic with its elements the numbers, worlds of algebra with variables, the world of Euclidean geometry with squares and triangles, topology with such objects as the Möbius strip and networks, fractals with objects that continually change — all are independent worlds yet are interrelated with one another. All form the universe of mathematics. A universe that can exist without anything from our universe, yet a universe that describes and explains things all around us.

Every mathematical world exists in a mathematical system. The system sets the ground rules for the existence of the objects in its world. It explains how its objects are formed, how they generate new objects, and how they are governed. A mathematical system is composed of basic elements, which are called *undefined terms*. These terms can be described, so that one has a feeling of what they mean, but technically they cannot be defined. Why? Because it takes terms to form definitions, and you have to begin with some terms. For these beginning words there are no other terms that exist which can be used to define them.

The best way to understand such a system is to look at one. Here's how a finite mini mathematical world might take form. Assume this mini world's undefined terms are *points* and *lines*. In addition to undefined terms, a mathematical system also has axioms, theorems, and definitions. *Axioms* (also called *postulates)* are ideas we accept as being true without proof. *Definitions* are new terms we describe/define using undefined terms or previously defined terms. *Theorems* are ideas which must be proven by using existing axioms, definitions or theorems.

What type of definitions, theorems and axioms can our mini world have? Here are some that might evolve—

> **Undefined terms:** *Points and lines.*
>
> **Definition 1:** *A set of points is collinear if a line contains the set.*
>
> **Definition 2:** *A set of points is noncollinear if a line cannot contain the set.*
>
> **Axiom 1:** *Our mini world contains only 3 distinct points, which do not lie on a line.*
>
> **Axiom 2:** *Any two distinct points make a line.*
>
> **Theorem 1:** *Only three distinct lines can exists in this world.*

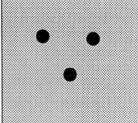

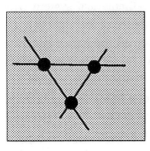

> *proof:*
> *Axiom 1 states that there are 3 distinct points in this world. Using Axiom 2 we know that every pair of these points determines a line. Hence three lines are formed by the three points of this world.*

This example illustrates how a mathematical world might evolve. As new ideas come to mind, one adds more undefined terms, axioms, definitions, and theorems and thereby expands the world.

The following sections introduce you to some mathematical worlds and their inhabitants.

GEOMETRIC WORLDS

*...The universe stands continually open to our gaze, but it cannot be understood unless one first learns to comprehend the language and interpret the characters in which it is written. It is written in the language of mathematics, and its characters are...geometric figures, without which it is humanly impossible to understand a single word of it; without these, one is wandering about in a dark labyrinth.—***Galileo**

Mathematics has many types of geometries. These include Euclidean and analytic geometries and a host of non-Euclidean geometries. Here we find hyperbolic, elliptic, projective, topological, fractal geometries. Each geometry forms a mathematical system with its own undefined terms, axioms, theorems and definitions. Although these geometric worlds may use the same names

This is an abstract design of Henri Poincaré's (1854-1912) hyperbolic world. Here a circle is the boundary of this world. The sizes of the inhabitants change in relation to their distance from the center. As they approach the center they grow, and as they move away from the center they shrink. Thus they will never reach the boundary, and for all purposes, their world is infinite to them.

for their elements or properties, their elements possess different characteristics. For example, in Euclidean geometry lines are straight and two distinct lines can either intersect in one point, be parallel, or be skew. But lines in elliptic geometry are not straight lines but great circles of a sphere, and therefore any two of its distinct lines always intersect in two points.

Consider the word *parallel*. In Euclidean geometry parallel lines are always equidistant and never intersect. Not so in elliptic or hyperbolic geometry. Why? Because every great circle of a sphere intersects another. Thus, elliptic geometry has no parallel lines. In hyperbolic geometry parallel lines never intersect, but they do not resemble Euclidean lines. Hyperbolic parallel lines continually come closer and closer together, yet never intersect. They

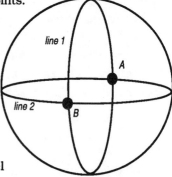

The above diagram shows two great circles, line 1 and 2 intersecting at points A & B.

are called asymptotic. Euclidean, hyperbolic, and elliptic geometries create three dramatically different worlds with lines and points, etc., but whose properties are universes apart. Each of these worlds is a mathematical system unto itself, and each has applications in our universe.

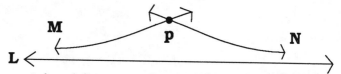

In hyperbolic geometry, lines M and N are both parallel to line L and pass through point P. M and N are asymptotic to line L.

NUMBER WORLDS

Numbers can be considered the first elements of mathematics. Their early symbols were probably marks drawn in the earth to indicate a number of things. But ever since mathematicians entered the scene the simple world of counting numbers has never been the same. Many people are familiar with integers,

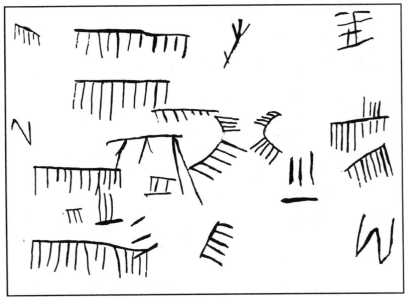

Stone Age number patterns found in La Pileta, Spain.

fractions and decimals, and use these for their daily computations. But number worlds also include the rational and irrational numbers, the complex numbers, the never-ending non-repeating decimals, transcendental numbers, transfinite numbers, and many many subsets of numbers that are linked by specific properties, such as perfect numbers whose proper factors total the number, or polygonal numbers whose shapes are connected to the shapes of regular polygons, and on and on. It is

interesting to delve into the interrelationship of numbers, surmise
how they developed, and explore their various properties.

The counting numbers date back to prehistoric times. Consider
the simple marks of the Stone Age number patterns from La Pileta
Cave in southern Spain, which was inhabited over 25,000 years
ago until the Bronze Age (1500 B.C.). The number π was known
over three thousand years ago, when it was used in the
calculations of a circle's area and circumference, and later shown
to be irrational and transcendental. Ancient civilizations were
aware that fractional quantities existed. The Egyptians used the
glyph for *mouth*, ⬤ , to write their fractions.

For example, ⬤ was 1/3, ⬤ was 1/10.

Irrational numbers were known by the ancient mathematicians,
who devised fascinating methods for approximating their values.
In fact, the Greeks developed the *ladder method* to approximatie
the $\sqrt{2}$ while the Babylonians used another method.

$0=$ two	$1=$ two	$10=$ two	$11=$ two	$100=$ two	$101=$ two
0	1	2	3	4	5

Hexagrams and their binary equivalents.

Over the centuries different civilizations developed symbols and
counting systems for numbers, and in the 20th century the binary
numbers and base two have been put to work with the computer
revolution. Gottfried Wilhelm Leibniz (1646-1716) first wrote
about the binary system in his paper *De Progressione Dyadica*
(1679). He corresponded with Père Joachim Bouvet, a Jesuit
missionary in China. It was through Bouvet that Leibniz learned
that the I Ching hexagrams were connected to his binary
numeration system. He noticed that if he replaced zero for each

broken line and 1 for the unbroken line, the hexagrams illustrated the binary numbers. Centuries prior to this, the Babylonians developed and improved upon the Sumerian sexagimal system to develop a base 60 number system. But this section on number worlds is not about number systems but about types of numbers.

$$1, 2, 3, 4, 5, 6, 7, 8, 9, 10 \ldots$$

Let's take a glimpse at the first type of numbers — the counting numbers. In the world of counting numbers we find the undefined terms are the numbers 1, 2, 3, ..., — with such axioms as *the order in which two counting numbers are added does not affect the sum (a+b=b+a, called the commutative property for addition); the order in which two counting numbers are multiplied together does not affect the product (axb=bxa, called the commutative property for multiplication).* — and such theorems as *An even number plus an even number is also an even number.* And, *The sum of any two odd numbers is always an even number.* But the world of counting numbers were not enough to solve all the problems that were to evolve over the years. Can you imagine tackling a problem whose solution was the value x for the equation x+5=3 and not knowing about negative numbers? What would have been some reactions — the problem is defective, there is no answer. Arab texts introduced negative numbers in Europe, but most mathematicians of the the 16th and 17th centuries were not willing to accept these numbers. Nicholas Chuquet (15th century) and Michael Stidel (16th century) referred to negative numbers as absurd. Although Jerome Cardan (1501-1576) gave negative numbers as solutions to equations, he considered them as impossible answers. Even Blaise Pascal said *"I have known those who could not understand that to take four from zero there remains zero."*

Thus history reveals that the solutions to certain problems required the invention of new numbers. For example, trying to explain the meaning of $\sqrt{-1}$ or to solve the equation $x^2 = -1$ led to the creation of imaginary numbers. And imaginary numbers led

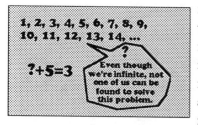

mathematicians to expand the world of numbers to include all real and imaginary numbers and more! Here's where the complex numbers (numbers of the form a+bi, where a and b are real and $i=\sqrt{-1}$) were introduced during the 16th century. With their introduction, all numbers invented thus far could be classified as complex.[1]

Graphs are a means of picturing mathematical objects. The real number line shows the the organization of all real numbers and their respective distances and sizes relative to zero. Every point on this line is linked to only one real number and vice versa— so -31/2 has only one location on this line. The imaginary numbers use an imaginary number line. So 2i is located as shown. Combining the real and imaginary number lines, we get a way to picture the complex numbers on what is called the complex number plane. The real and imaginary lines are perpendicular at their origins. Every point in the plane they form is associated with only one complex number — no other complex number has that location. So the ordered pair for the point (-4, 3) is matched with the complex number -4 +3i. (4,0) means 4+0i which equals the number 4. This coordinate system was an ingenious way to organize and picture the complex numbers. The questions now is — Are there any numbers which do not appear on this plane? You bet! The chapter *The Magic of Numbers* has some examples.

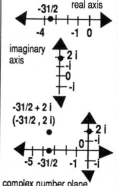

[1] Any real number can be thought of as a complex number whose imaginary part is 0, and any imaginary number is a complex number with real part 0.

THE WORLDS OF DIMENSIONS

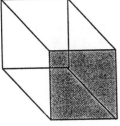

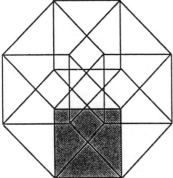

The four dimensions

Let's look at the worlds which are created by the idea of dimensions. A mathematical world can exist on a single point, on a single line, on a plane, in space, in a hypercube (tesseract). Each higher dimension encompasses those beneath it, yet each lower dimension can be a world in itself. Imagine your world and your life on a flat plane. You cannot look up or down. Three dimensional creatures can invade your world without you even knowing by simply entering your domain from above or below. Mathematicians, writers, and artists have used various ideas to try to capture the essence of different dimensions in their works. Dimensions beyond the third have alway been intriguing. The cube was one of the first 3-D objects to be introduced into the fourth dimension by becoming a hypercube. The stages for arriving at a hypercube are illustrated . Computer programs have even been devised to derive glimpses of the fourth dimension by picturing 3-D perspectives of the various facets of the hypercube.

THE WORLDS OF INFINITIES

*To see the world in a
grain of sand,
And a heaven in a
wildflower;
Hold infinity in the palm of your hand,
And eternity in an hour.* —**William Blake**

Infinity has stimulated imaginations for thousands of years. It is an idea drawn upon by theologians, poets, artists, philosophers, writers, scientists, mathematicians — an idea that has perplexed and intrigued — an idea that remains illusive. Infinity has taken on different identities in different fields of thought. In early times, the idea of infinity was, rightly or wrongly, linked to large numbers. People of antiquity experienced a feeling of the infinite by gazing at stars and planets or at grains of sand on a beach. Ancient philosophers and mathematicians such as Zeno, Anaxagoras, Democritus, Aristotle, Archimedes pondered, posed and argued the ideas that infinity presented.

Aristotle proposed the ideas of potential and actual infinities. He argued that only potential infinity existed.[1]

In *The Sand Reckoner* Archimedes dispelled the idea that the number of grains of sand on a beach are infinite by actually determining a method for calculating the number on all the beaches of the earth.

Infinity has been the culprit in many paradoxes. Zeno's paradoxes of Achilles and the tortoise and the Dichotomy[2] have perplexed readers for centuries. Galileo's paradoxes[3] dealing with segments, points, and infinite sets should also be noted.

This field of sunflowers in the Spanish countryside gives the illusion of infinity.

The list of mathematicians with their discoveries and uses or misuses of infinity extends through the centuries. Euclid (circa 300 B.C.) showed that the prime numbers were infinite by showing there was no last prime. Headway in the realm of the infinite was made by Bernhard Bolzano (1781-1848), Gottfried W. Leibniz (1646-1716), and J.W.R. Dedekind (1831-1916). But the phenomenal work of Georg Cantor (1845-1918) on set theory was a major breakthrough. Building, creating and refining ideas, Cantor found a new way to organize mathematics by use of the notion of a set. He determined a way to compare infinite sets by developing transfinite numbers — numbers that dared to cross the realm of the finite. Using the idea of equivalent sets and countability, he determined which infinite sets had the same number of objects and assigned them a transfinite number. His work and proofs on these topics are ingenious.

In addition to teasing our minds, infinity is an indispensable mathematical tool. It has played a crucial role in many mathematical discoveries. We find it used in: determining the areas and volumes both in geometry and calculus — calculating approximations for π, and e and other irrational numbers— trigonometry— calculus — half-lives — infinite sets — self-perpetuating geometric objects — limits — series — dynamic symmetry — and more. Other parts of this book explore various notions of infinity, such ideas as infinite generating fractals, the chaos theory, the continual search for a larger prime number, transfinite numbers and others *ad infinitum.*

---

[1]The counting numbers are potentially infinite, since one can be added to any number to get the next, but the entire set cannot be actually attained.

[2] In *The Dichotomy Paradox* Zeno argues that a traveler walking to a specific destination will never reach the destination because the traveler must first walk half the distance. Reaching this halfway point, the traveler then has to walk half the remaining distance. Then half of the part that remains. Since there will always be half of the part that remains to walk and an infinite number of halfway points to pass, the traveler will never reach the destination.

[3] In Galileo's 1634 work, *Dialogues Concerning Two New Sciences,* he discusses infinity in relation to the positive integers and the squares of the positive integers. He even deals with one-to-one correspondence between these two infinite sets. But he reaches the conclusion that the concepts of *equality, greater than,* and *less than* were only applicable to finite sets. Galileo believed the principle that *the whole is always greater than its parts* had to apply to both finite and infinite sets. Three hundred years later, Cantor showed this principle did not hold for infinite sets and used the idea of one-to-one correspondence to revise the traditional notions of *equality, greater than,* and *less than* when dealing with infinite sets. Cantor's modifications did away with many paradoxes involving *infinite sets* and *the whole is always greater than its parts.*

FRACTAL WORLDS

I coined fractal from the Latin adjective fractus. The corresponding Latin verb fragere means 'to break': to create irregular fragments. ... how appropriate for our needs! — that, in addition to 'fragmented' (as in fraction or refraction) fractus should also mean 'irregular', both meanings preserved in fragment. —**Benoit Mandelbrot**

Fractals are magnificent objects which come in infinitely many shapes. Ernesto Cesaro (Italian mathematician 1859-1906) wrote this about the geometric fractal, the Koch snowflake curve— *What strikes me above all about the curve is that any part is similar to the whole. To try to imagine it as completely as possible, it must be realized that each small triangle in the construction contains the whole shape reduced by an appropriate factor. And this contains a reduced version of each small triangle which in turn contains the*

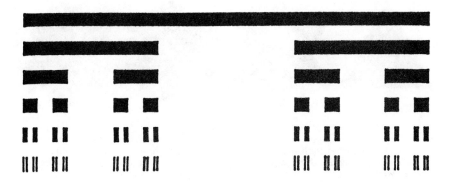

In 1883, Cantor constructed this fractal called the Cantor set. Starting with the segment of length the unit interval on the number line, Cantor removed the middle one third and got stage 1. Then to each remaining 3rds, removed the middle one-third, thereby creating the 2nd stage. Repeating the process ad infinitum, the infinite set of points that remains is called the Cantor set. Here are the first stages of the Cantor set.

whole shape reduced even further and so on to infinity... It is this self-similarity in all its part, however small, that makes the curve seems so wondrous. If it appeared in reality it would not be possible to destroy it without removing it altogether, for otherwise it would cease-lessly rise up again from the depths of its triangles like the life of the universe

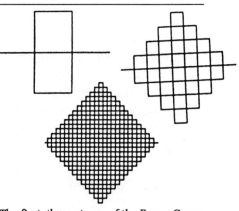

The first three stages of the Peano Curve. The Peano curve was made in the 1890's, by repeatedly applying successive generation to a segment.

itself.   This is the essence of fractals. If a portion of it remains, that portion retains the essence of the fractal—which in turn can regenerate itself. So what is a fractal? Perhaps mathematicians have purposely avoided giving a definition to not restrict or inhibit

The first four stages of the Koch snowflake. The Koch snowflake is gener-ated by starting with an equilateral triangle. Divide each side into thirds, delete the middle third, and construct a point off that length out from the deleted side.

the creativity of fractal creations and ideas that are formulating in this very new field of mathematics. With this new field, ideas such as fractional dimensions, iteration theory, turbulence applica-

The first four stages of the Sierpinksi tri-angle. Begin with an equilateral triangle. Divide it into four congruent triangles as shown and remove the middle one. Repeat this process to the smaller triangles formed ad infinitum. The resulting fractal has in-finite perimeter and zero area!

tions, self-similarity have evolved. Applications for fractals range from acid rain to zeolites, from astronomy to medicine, from cinematography to cartography to economics, and on and on.

Mathematically speaking, a fractal is a form which begins with an object — such as a segment, a point, a triangle— that is constantly being altered by reapplying a rule ad infinitum. The rule can be described by a mathematical formula or by words. The previous diagrams illustrate four of the earliest fractals made.

One can think of a fractal as an ever growing curve. To view a fractal, you must really view it in motion. It is constantly developing. Today we are fortunate to have computers capable of generating fractals before our eyes. It was equally fortunate that Benoit Mandelbrot, in the same spirit of the early mathematicians, studied and expanded the ideas and applications of fractals almost singlehandedly from 1951-75. In fact, he coined the word *fractal.* How astonished the adventurous mathematicians[1] of the 19th century, who first dared to look at these ideas most considered monstrous[2] and psychopathic, would be to see the wondrous geometry of fractals in motion.

When we view an illustration or photograph of a fractal, we are seeing it for one moment in time — it is frozen at a particular stage of its growth. In essence it is this idea of growth or change that links fractals dramatically to nature. For what is there in nature that is not changing? Even a rock is changing on a molecular level. Fractals can be designed to simulate almost any shape you can imagine. Fractals are not necessarily confined to one rule, but a series of rules and stipulations can be the rule. Try creating your own fractal. Pick a simple object and design a rule to apply to it.

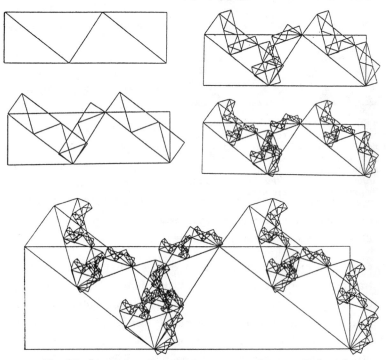

The first five stages of a computer generated geometric fractal

[1] Mathematicians Georg Cantor, Helge von Koch, Karl Weierstrass, Dubois Reymond, Guiseppe Peano, Waclaw Sierpinski, Felix Haussdorff, A.S. Besicovitch (Haussdorff and Besicovitch worked on fractional dimensions), Gaston Julia , Pierre Fatou (Julia and Fatou worked on iteration theory), Lewis Richardson (worked on turbulence and self-similarity) — spanning the years from 1860's to early 20th century— explored ideas dealing with the "monsters".

[2] These "monsters" were neither accepted or considered worth exploring by conservative mathematicians of the time. It was felt that fractals contradicted accepted mathematics because some were continuous functions that were not differentiable, some had finite areas and infinite perimeters, and some could completely fill space.

THE PARABLE OF THE FRACTAL

"Wake-up Fractal! You must get to work," the voice prodded the sleeping Fractal.

"Not again and so early," pleaded Fractal. "I just got my dimensions in order."

"Wake-up Fractal! Come down from that cloud you made," the voice prodded the sleeping Fractal.

"You're needed at the Geological Survey—another coastline needs to be described," the voice continued.

"When will I get a break?" questioned Fractal

"You've had it easy for centuries — now it's time to get to work," the voice replied.

"Work, work, work. Why don't they call on Square, Circle, Polygon, or any other Euclidean figure? Why me?" asked Fractal.

"First you complained at being ignored, and that they called you a monster. Now that they're finally understanding you, you want to retire. Just be thankful you are so popular," the voice rebutted.

"Popular is one thing, but they won't let me rest. It's never been the same since that Mandelbrot christened me and gave me my debut," replied Fractal. "Mathematicians were tediously struggling with me. I'm sure my fractional dimensions threw them off for a while. Those poor souls from the 19th century had no computers to help them. Most mathematicians would not accept me, for I did not fit or follow their mathematical rules. But some mathematicians were stubborn. Now here I am, being designed and used in so many areas—computers certainly were a boon. One moment the screen displays a fragment or beginning part of a fractal and the next moment the screen is being filled with its generations— ever growing. They are now using me in almost everything— I can describe roots, vegetables, trees, popcorn, clouds, scenery . . . I must say its is very exciting to stretch my limits. I love to do coastlines because it still baffles many people to learn I can enclosed a region whose area is finite while my perimeter is infinite. I'm serving for modeling many of the world's phenomena. For example— population with Peano curves, fractal curves for creating scene in movies, fractals for describing astronomy, meteorology, economics, ecology, et cetera, et cetera, et cetera. I'm so busy and involved that things are beginning to get a bit chaotic, especially since they mixed me into the chaos theory," Fractal said sounding very tired.

The voice started again. "Stop complaining! Chaos theory offers you some variety. Without it you'd just be continually repeating the same rule and generating the same old shape over and over, at

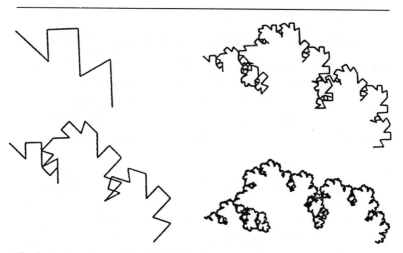

The beginning stages of a fractal cloud.

least when some initial input is slightly varied, something totally different can evolve."

"I suppose you're right, " Fractal sighed.

"Of course I'm right. Just think how boring it would be to be the same shape forever, like a poor square or a circle," the voice asserted.

"Well, at least there are no surprises for a square or circle." Fractal countered.

"That's precisely it. Life is full of surprises; that's why they are calling on you so often. You are more like life. " The voice seemed complementing Fractal.

"You mean I'm human?" Fractal asked.

"I wouldn't go that far. And besides all life isn't human. Let's say you're just different , and you're non-Euclidean!" And with that comment the voice drifted off.

FINDING THE AREA OF A SNOWFLAKE CURVE

This beautiful geometric fractal was created in 1904 by Helge von Koch. To generate a Koch snowflake curve, begin with an equilateral triangle. Divide each side into thirds. Delete the middle third, and construct a point of that length out from the deleted side. Repeat the process for each resulting point ad infinitum.

Two fascinating properties, which seem contradictory, are—
•the area of the snowflake curve is 8/5 of the area of the original triangle that generates it; •the perimeter of the snowflake curve is infinite.

Here is an informal proof that the area of the snowflake curve is 8/5 of its generating triangle.

I. Assume the area of equilateral ΔABC is k.

II. Divide ΔABC into nine congruent equilateral triangles of area , a , as shown. Thus k=9a.

Now concentrate on determining the limit of the area of one of the 6 initial points of the snowflake curve. We know the area of the large point is a, since its one of the nine triangles. The next set of points generated from it have area (a)(1/9) each, just like the original triangle had been divided into 9 congruent triangles it also is. In fact, each successive point is broken down into nine congruent triangles with two triangles springing from it.

STEP III shows the summation of the various areas of this point.

STEP IV: Now, by adding up the areas created by each of the 6 points plus the hexagon in the interior of the original generating triangle, we get expression IV.

STEP IV is changed to **STEP V.** The resulting series in the

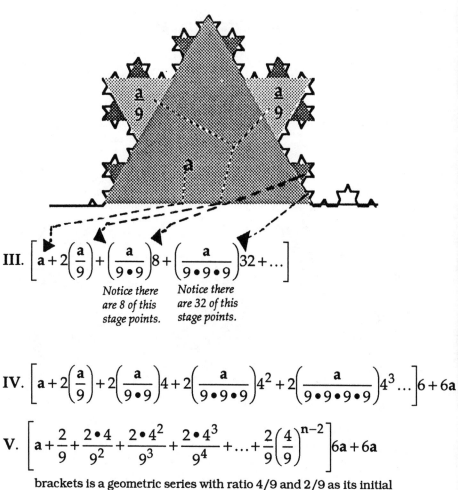

III. $\left[a + 2\left(\dfrac{a}{9}\right) + \left(\dfrac{a}{9\bullet9}\right)8 + \left(\dfrac{a}{9\bullet9\bullet9}\right)32 + \ldots\right]$

Notice there are 8 of this stage points.   *Notice there are 32 of this stage points.*

IV. $\left[a + 2\left(\dfrac{a}{9}\right) + 2\left(\dfrac{a}{9\bullet9}\right)4 + 2\left(\dfrac{a}{9\bullet9\bullet9}\right)4^2 + 2\left(\dfrac{a}{9\bullet9\bullet9\bullet9}\right)4^3 \ldots\right]6 + 6a$

V. $\left[a + \dfrac{2}{9} + \dfrac{2\bullet4}{9^2} + \dfrac{2\bullet4^2}{9^3} + \dfrac{2\bullet4^3}{9^4} + \ldots + \dfrac{2}{9}\left(\dfrac{4}{9}\right)^{n-2}\right]6a + 6a$

brackets is a geometric series with ratio 4/9 and 2/9 as its initial term, so we can calculate its limit. $(2/9)/(1-(4/9))=2/5$.

STEP VI. Substituting the 2/5 for the limit of the series, we get

[1+2/5]6a+6a = 72a/5

Now we need to express the area of the snowflake curve in terms of **k**, the area of the original generating triangle. Since **k=9a**, we get **a=k/9**. Substituting this for **a** in **72a/5**, we get **(72/5)(k/9)** = **(8/5)k.**

MONSTER CURVES

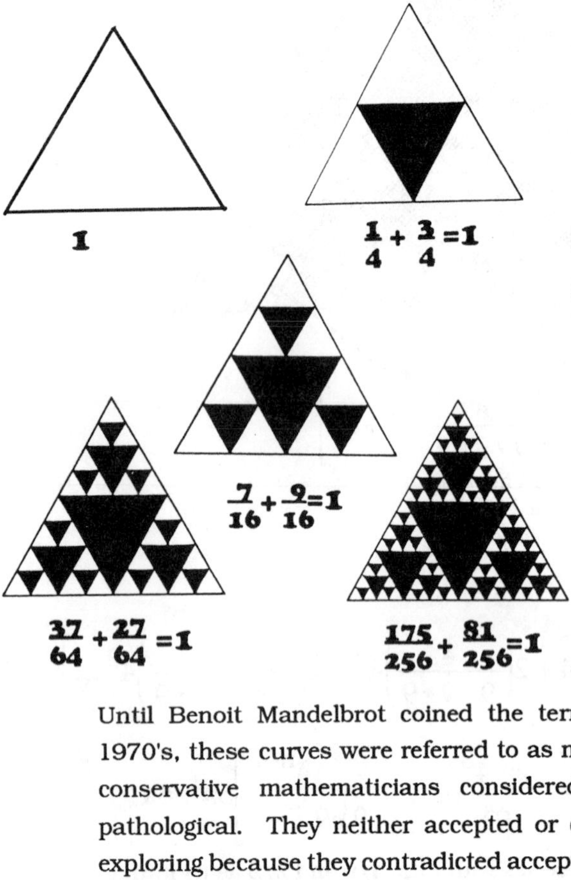

The stages of the Sierpinski triangle . Suppose the area of the initial generating equilateral triangle is 1 square unit. The sum of the areas of the black and white triangles are indicated through the first five generations. Suppose the black triangle represents removal of area. Notice how the value for the white triangles is continually decreasing, meaning the white area is approaching zero. Thus the area for the Sierpinski triangle approaches 0, while its perimeter approaches infinity.

1

$$\frac{1}{4} + \frac{3}{4} = 1$$

$$\frac{7}{16} + \frac{9}{16} = 1$$

$$\frac{37}{64} + \frac{27}{64} = 1$$

$$\frac{175}{256} + \frac{81}{256} = 1$$

Until Benoit Mandelbrot coined the term "fractal" in the late 1970's, these curves were referred to as monsters. 19th century conservative mathematicians considered these monster curves pathological. They neither accepted or considered them worth exploring because they contradicted accepted mathematical ideas. For example, some were continuous functions (functions without any gaps) that were not differentiable, some had finite areas and infinite perimeters, and some could completely fill space. The *Sierpinski triangle* [1] (also called *Sierpinski gasket*) has an infinite perimeter and a finite area. The illustration above tries to point out how the *Sierpinski triangle* 's area is zero.

[1]Named after mathematician Waclaw Sierpinski (1882-1920).

MANDELBROT SET CONTROVERSY

In the 17th century a number of prominent mathematicians (Galileo, Pascal, Torricelli, Descartes, Fermat, Wren, Wallis, Huygens, Johann Bernoulli, Leibniz, Newton) were intent on discovering the properties of the cycloid. Even as there were many discoveries at this period of time, there were also many arguments about who had discovered what first, accusations of plagiarism, and minimization of one another's work. As a result, the cycloid has been labeled *the apple of discord* and the *Helen of geometry*. 20th century mathematicians now seem to have a new Helen of geometry— the Mandelbrot set.

Who first discovered the Mandelbrot set[1]? This is a very heated question among present day mathematicians. The contenders are:

— *Benoit Mandelbrot is often described as a pioneer because of his initial work on fractals in the 1970s. Mandelbrot's work showing variants of the Mandelbrot set was published December 26, 1980 in Annals of the New York Academy of Sciences. His work on the actual Mandelbrot set was published in 1982.*

— *John H. Hubbard of Cornell University and Adrien Douady of the University of Paris named the set Mandelbrot in the 1980s while working on proofs of various aspects of the set. In 1979, Hubbard says he met with Mandelbrot, and showed Mandelbrot how to program a computer to plot iterative functions. Hubbard admits that Mandelbrot later developed a superior method for generating the images of the set.*

— *Robert Brooks and J. Peter Matelski claim they independently discovered and described the set prior to Mandelbrot, although their work was not published until 1981.*

— *Pierre Fatou described Julia sets' unusual properties around 1906, and Gaston Julia's work on Julia sets predates Fatou's. (Julia sets acted as springboards for Mandelbrot sets.)*

Who gets the credit? Perhaps all.

---

[1]The illustration above is the most familiar fractal form from the Mandelbrot set. The Mandelbrot set is a treasure trove of fractals, which contains an infinite number of fractals. The set is generated by an iterative equation, e.g. z^2+c, where z and c are complex numbers and c produces values than are confined to a certain boundary.

MATHEMATICAL WORLDS IN LITERATURE

*There is an astonishing
imagination even in
the science of mathematics.*

— Voltaire

Is the tesseract the figment of a mathematical imagination? Is the only "real" dimension the 3rd dimension? We learn in Euclidean geometry that a point only shows location, and it cannot be seen since it has zero dimension. Yet we can see a line segment composed of these invisible points. A line is infinite in length, yet does such a figure exist in the realm of our lives? What about a plane? Infinite in two dimensions and only one point thick. What is a plane in our world? Consider the pseudosphere of hyperbolic geometry; asymptotic lines of exponential functions, infinities of transfinite numbers. Consider the imaginary numbers, the complex number plane, fractals, and even a circle. *One wonders if these can exist in our world.* Although there is no doubt of their existence in their respective mathematical systems, these concepts are only models in our world.

Many writers, artists and mathematicians have ingeniously used these concepts to describe worlds where these ideas come to life. Such writers as Dante, Italo Calvino, Jorge Luis Borges, and Madeleine L'Engle have drawn on mathematics to enhance their creations.

In the 19th century, mathematician Henri Poincaré created a model of a hyperbolic world contained in the interior of a circle. Here to all things and inhabitants, their circular world was infinite. Unbeknownst to these creatures, everything would shrink as it moved away from the center of the circle, while growing as it approached the center. This meant the circle's boundary was never to be reached, and hence their world appeared infinite to

Circle Limit IV (Heaven & Hell) *by M.C. Escher depicts a world reminiscent of Henri Poincarè's hyperbolic world.*

them. In 1958, artist M.C. Escher created a series of woodcuts, entitled *Circle Limit* I, II, III, IV which convey a feeling of what Poincaré had described. Escher described a world as *"the beauty of this infinite world-in-an-enclosed plane."* [1]

For her novel, *A Wrinkle in Time,* Madeleine L'Engle uses the tesseract and multiple dimensions as means of allowing her characters to travel through outer space. *"...for the 5th dimension you'd square the fourth and add that to the other four dimensions and you can travel through space without having to go the long way* around...*In other words a straight line is not the shortest distance between two points."*

Italo Cavino describes a world that exists in a single point in his short story *All At One Point*. His ingenious creativity makes one believe such a zero dimensional world actually exists. *"Naturally, we were all there, — Old Qfwfq said,— where else could we have been? Nobody knew then that there could be space. Or time either: what use did we have for time, packed in there like sardines? I say "packed like sardines," using a literary image: in reality there wasn't even space to pack us into. Every point of each of us coincided with every point of each of the others in a single point, which was where we all were."*

Looking back to the Middle Ages and Dante's *The Divine Comedy*, we find Euclidean geometric objects were the bases for Dante's hell. The cone's shape was used to hold people in stages of hell. Within it, Dante had nine circular cross-sections that acted as platforms which grouped people by sins committed.

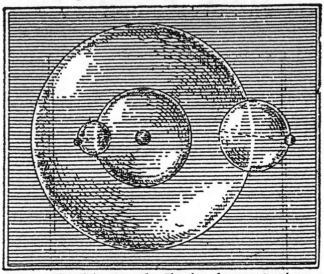

From Dante's **The Divine Comedy.** *The plan of concentric spheres, which shows the Earth in the sphere (bearing the epicycle) of the Moon, and these are also enclosed in the sphere (bearing the epicycle) of Mercury.*

In the 1900's infinity was featured in Jorge Luis Borges' *The Book of Sand.* Here the main character acquires a "marvelous" book. *"The number of pages in this book is no more or less than infinite. None is the first page, none the last. I don't know why they're numbered in this arbitrary way. Perhaps to suggest the terms of an infinite series admit any number."* This book adversely changes his life and his outlook on things, until he realizes he must find a way to dispose of it— *"I thought of fire, but I feared that the burning of an infinite book might likewise prove infinite and suffocate the planet with smoke."* What would your solution be? You might want to read the book to find how the hero resolved his dilemma.

Science fiction writers have utilized mathematical ideas to help create their worlds. For example, in an episode of *Star Trek — The Next Generation,* the starship is being pulled by an "invisible" force toward a black hole. Only when the ship's schematic monitor changes perspective does the crew realize the unknown force is a 2-dimensional world of minute life forms.

Mathematics is full of ideas that make one's imagination churn and wonder— *Are they real?* To mathematicians they are real. Mathematicians are familiar with the worlds in which these ideas reside — perhaps not within our realm, but real in their own nonetheless!

---

[1] *M.C.Escher,* Harry N. Abrams,Inc., New York, 1983.

MATHEMATICS & ART

ART, THE 4TH DIMENSION & NON-PERIODIC TILING

MATHEMATICS & SCULPTURE

MATHEMATICAL DESIGNS & ART

MATHEMATICS & THE ART OF M.C.ESCHER

PROJECTIVE GEOMETRY & ART

MIXING MATHEMATICS & ART OF ABRECHT DÜRER

COMPUTER ART

> *The most beautiful thing we can experience is the mysterious. It is the source of all true art and science.*
> **—Albert Einstein**

Linking mathematics and art may seem alien to many people. But mathematical worlds of geometries, algebra, dimensions, computers have provided tools for artists to explore, enhance, simplify, and perfect their work. Over the centuries, artists and their works have been influenced by the knowledge and use of mathematics. The ancient Greek sculptor, Phidias, is said to have used the golden mean in the proportions of many of his works. Albrecht Dürer employed concepts from projective geometry to achieve perspective, and geometric constructions played a vital role in his typography of Roman letters. Since religious doctrine prohibited the use of animate objects in

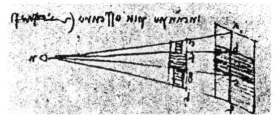

A sketch from one of Leonardo da Vinci's notebooks illustrating lines converging to a vanishing point.

Moslem art, Moslem artists had to rely on mathematics as an avenue for their artistic expression, thus leading them to create a wealth of tessellation designs. Leonardo da Vinci felt "...*no human inquiry can be called science unless it pursues its path through mathematical exposition and demonstration.*" Leonardo's sculptures and paintings illustrate his study of the golden rectangle, proportions, and projective geometry, while his architectural designs show his work in geometric structures and knowledge of symmetry. The topics in this section are a few examples illustrating the connection between mathematics and art.

Mathematics takes us into the region of absolute necessity, to which not only the actual world, but every possible world, must conform. —**Albert Einstein**

ART, THE 4TH DIMENSION & NON-PERIODIC TILING

On a canvas, the artist is restricted to two-dimensions to communicate the feeling of other dimensions. Icon artists of the Byzantine period depicted three-dimensional religious scenes in only two-dimensions, giving the subject matter a mystical appearance. During the Renaissance, artists using the concepts of projective geometry transformed their flat canvas into the three-dimensional world they wanted to convey. Today, mathematics plays an active role in providing inspiration and tools for the creation and communication of an artist's ideas. Artists use mathematical ideas to escape into higher dimensions. The hypercube[1], for example, has been used by artists to take a step into the fourth-dimension. In the early 1900's architect Claude Bragdon adapted the hypercube along with other four-dimensional designs in his work[2]. Intrigued by the hypercube, Salvador Dali[3] delved into mathematics for his model of an unfolded hypercube which is the focal point in his painting *The Crucifixion*[4] (1954).

Today, there are a number of artists pursuing art in connection with mathematical ideas — in particular, mathematics of non-periodic tiling, multi-dimensions and computer renditions. In fact, computer renditions of the hypercube, created by mathematician Thomas Banchoff and computer scientist Charles Strauss of Brown University, produce visualization of the hypercube moving in and out of a 3-D space. Various images of the hypercube in the 3-D world are thereby captured on the computer

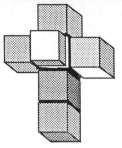

The unfolded hypercube was the inspiration for Salvador Dali's *The Crucifixion* (1954). Metropolitan Museum of Art, Gift of the Chester Dale Collection, 1955. (55.5)

monitor. Introduced to this part of mathematics, artist Tony Robbin has created 3-D canvas representations of the hypercube with the canvas acting as a plane intersecting the hypercube. Nonperiodic tilings, Penrose tiles, quasicrystal geometry and fivefold symmetry[5], have aided Robbin in creating fascinating structures, which change dramatically according to the perspective of the viewer. One moment one views a series of triangles while in the next position interlaced pentagonal stars appear. The combination of non-periodic tiling of both 2 and 3-D forms intertwined in an unusual type of symmetry create an almost contradictory image.

---

[1] Also known as the tesseract—a 4-dimensional representation of a cube.

[2] At the same time Bragdon used magic lines in architectural ornaments and graphic designs of books and textiles.

[3] Dali contacted the mathematics department at Brown University for further information.

[4] Jesus Christ is nailed to a cross represented by the unfolded fourth-dimensional hypercube.

[5]*Non-periodic tiling* is tessellating with tiles or shapes which create designs which have no pattern.

n-fold symmetry : If a pattern is preserved when rotated 360°/n, it is said to have n-fold symmetry. Therefore, a pattern has *fivefold symmetry* if a rotation of 72° retains the pattern.

Quasicrystals are a newly discovered state of solid matter. Solid matter was thought to exist only in two states, amorphorous or crystalline. In amorphorous the atoms (or molecules) are arranged randomly, while in crystalline the arrangement is the periodic repetition of a unit cell building block. The discovery of quasicrystals revealed a new state in which the arrangement of units is non-periodic and with an unusual symmetry, e.g. fivefold, not present in amorphorous or crystalline matter.

MATHEMATICS & SCULPTURE

Dimensions, space, center of gravity, symmetry, geometric objects, and complementary sets are all mathematical ideas which come into play when a sculptor creates. Space plays a prominent role in a sculptor's works. Some works simply occupy space in the same way we and other living things do. In these works the center of gravity[1] is a point within the sculpture. These are objects that are anchored to the ground and occupy space in a manner with which we are comfortable or accustomed. For example Michelangelo's *David*, the *Discobolus* by the ancient Greek artist, Myron, or Beniamino Bufano's *St. Francis on Horseback* all have their center of gravity within the mass of their sculpture. Some modern art sculptures play with space and its three dimensions in unconventional ways. These use space as an integral part of the work. Consequently the center of gravity can be a point of space rather than a point of the mass, as illustrated by such works as *Red Cube* by Isamu Noguchi, the *Eclipse* by Charles Perry, and the *Vaillancourt Fountain*

The Discobolus (circa 450 B.C.) by Myron, cast in bronze, captures a moment in motion.

by Louis Vaillancourt. Other sculptures depend on their interaction with space. Here the space around the artwork (the complementary set of points of the mass) is as, or equally, important as the sculpture. Consider *Zinc Zinc Plain* by Carl Andre. This sculpture is staged in a room devoid of any other works or objects. The plane is created by 36 small squares forming a square which lies flat on the floor. The room represents space, the set of

San Francisco's controversial *Vaillancourt Fountain* has as its center of gravity a point of space.

all points, and he describes his work as "a cut of space".[2] Some works seem to defy gravity. These include such sculptures as the mobiles of Alexander Calder with their exquisite balance and symmetry and Isamu's Noguchi's *Red Cube* balancing mysteriously on its vertex. There are even sculptures which use the Earth itself as an integral part of the art and its statement, e.g. *The Running Fence* by Cristo, Carl Andre's *Secant,* and the mysterious geometric grass theorems appearing in England.

Often the physical nature of an artist's conceptional work requires mathematical understanding and knowledge to make the work possible. Leonardo da Vinci mathematically analyzed most of his creations before undertaking a work.   If M.C. Escher had not mathematically dissected the ideas of tessellation and optical illusions, his works would not have evolved with the ease with which he was able to undertake them once he understood the mathematics of these ideas.

Today there are many examples of sculptors looking at mathematical ideas to expand their art. Tony Robbin uses the study of quasicrystal geometry, 4th dimensional geometry, and computer science to develop and expand his art. In his *Easter egg* giant sculpture Ronald Dale Resch had to use intuition, ingenuity, mathematics,and the computer as

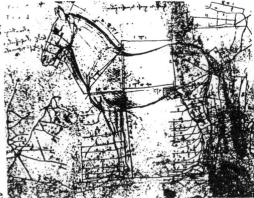

This sketch by Leonardo shows his analysis of the horse's anatomy.

Author in front of *Continuum* by Charles Perry. National Air & Space Museum, Washington D.C.

well as his hands to complete his creation. And artist-mathematician Helaman R.P. Ferguson uses traditional sculpting, the computer and mathematical equations to create such works as *Wide Sphere* and *Klein Bottle with Cross-cap & Vector*. Consequently it is not surprising to find mathematical models doubling as artistic models. Among these we find the cube, the polycube, the sphere, the torus, the trefoil knot, the Möbius strip, polyhedrons, the hemisphere, knots, squares, circles, triangles, pyramids, prisms,.... Mathematical objects from Euclidean geometry and topology have played important roles in the sculptures of such artists as Isamu Noguchi, David Smith, Henry Moore, Sol LeWitt.

An Alexander Calder mobile. East Building of the National Gallery of Art, Washington, D.C. .

Regardless of the sculpture, mathematics is inherent in it. It may have been conceived and created without a mathematical thought, nevertheless mathematics exists in that work, just as it exists in natural creations.

---

[1]The center of gravity is the point on which an object can be balanced. For example, the center of gravity or centroid of a triangle can be determined by drawing that triangle's medians. The point where the three medians intersect happens to be the center of gravity.

[2]Art & Physics, Leonard Shlain, William Morrow & Co. NY, 1981.

PUTTING MATHEMATICS INTO STONE

Trefoil knots — torus — spheres — vectors — flow — movement — these are some of the mathematical ideas inherent in the sculptures of Helaman Ferguson.

We have often heard of artists using mathematical ideas to enhance their work. Mathematician- artist Helaman R.P. Ferguson conveys the beauty of mathematics in his phenomenal sculptures. As he states "Mathematics is both an art form and a science...I believe it is feasible to communicate mathematics along aesthetic channels to the general audience."[1]

Eine Kleine Rock Musik III
Photography by Ed Bernik. From *Helaman Ferguson: Mathematics in Stone and Bronze* by Claire Ferguson. Meridian Creative Group, Copyright © 1994.

To create his exquisite forms, Helaman utilizes methods from traditional sculpting, the computer, and mathematical equations. His works bear such names as *Wild Sphere; Klein Bottle With Cross-cap And Vector Field, Torus, Umbilical Torus With Vector Field., Whaledream II* (horned sphere).

---

[1] Ivars Peterson. *Equations in Stone, Science News* Vol. 138 September 8, 1990.

LAYING AN EGG MATHEMATICALLY

When Ronald Dale Resch accepted the commission to design a gigantic Easter egg sculpture for Vegreville, Alberta, he soon discovered he would have to develop the mathematics for the task virtually from scratch.

Over the years Resch has refined the art of manipulating 2-D objects into 3-D forms. His work and the problems he has solved point to mathematics, yet he has had little formal mathematical training. Working with sheets of various materials such as paper or aluminum, he transforms them into works of art by folding techniques he has developed. He solves geometric problems using intuition, ingenuity, mathematics, the computer and his hands.

His initial instincts about the design of the egg were that he could make two ellipsoids for the ends and a bulging cylinder for the center. He quickly realized this would not work. Discovering that available mathematics for the egg was limited[1], he realized he would have to go it alone. His resulting design for the magnificent Easter egg required 2,208 identical equilateral triangular tiles and 524 three-pointed stars tiles, which were equilateral and whose width varied according to their location on the egg. He found that by varying the angle of placement of the tiles ever so slightly (from less than 1˚ to 7˚), the flat tiles gave the impression of curving and the contour of the egg resulted. The final structure is 25.7 feet long and 18.3 feet wide, and weighs 5000 pounds.

[1]Algebraic equations for egg-shaped curves were developed by French mathematician René Descartes (1596-1650). In 1842, as a youth, Scottish mathematician James Clerk Maxwell (1831-1879) devised a method for constructing an egg using a pencil, string and tacks.

MATHEMATICS DESIGNS & ART

The following figures can and have been used to create artistic and graphic designs. Their bases are mathematical, but that does not take away from their beauty and simple elegance.

MATHEMATICAL STARS

The illustrations show how to generate stars by using regular polygons. For odd sided polygons simply join every other vertex of the polygon, and you will arrive back at the starting point. For even pointed stars, note that the star is made up of two rotated polygons. These will have half the number of sides as the generating polygon.

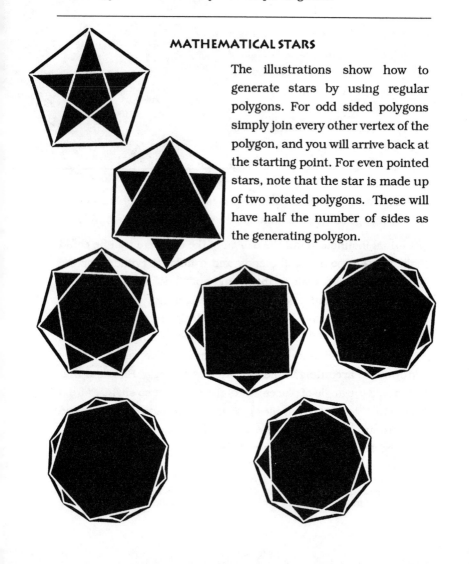

MATHEMATICAL EMBROIDERY

Math enthusiasts have been "embroidering" mathematical curves over the centuries. It is always fascinating to discover a curve formed from a series of straight line segments. The "stitches" (line segments) end up being tangents for the curve they form. One can mathematically "embroider" a circle, an ellipse, a parabola, a hyperbola, a cardioid, a limaçon, a deltoid, etc. In the process of embroidering these curves, one discovers some of their special characteristics. Here is how an astroid can be made.

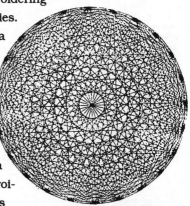

Concentric circles appear when the diagonals of a 24-sided polygon are stitched.

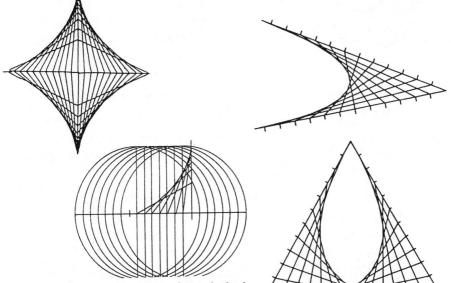

One way to make an astroid is to think of it as a sliding ladder. Circles, the radius of the ladder, are used to mark the base of the ladder's new location.

Variations of parabolas.

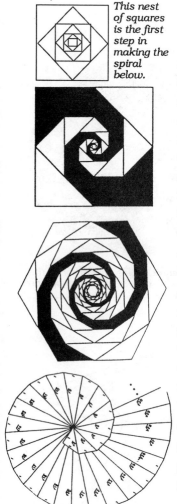

This nest of squares is the first step in making the spiral below.

STRAIGHT LINE SPIRALS

Spirals are mathematical objects that seem to convey motion. They encompass a family of curves from two-dimensional spirals — equiangular (logarithmic) and Archimedean spirals for examples — to three-dimensional spirals such as the conchoid and helix spirals. Spirals touch many areas in our lives and many living objects. A few examples are — antlers of certain goats, seedheads of flowers, many sea shells, the growth of certain vines, DNA, architecture, art, graphic designs.

Straight line spirals are formed by making a nest of regular polygons in which each reduced polygon is formed by connecting the midpoints of its sides. Looking at a nest of polygons, the spiral is hard to see until it is shaded. Here are some exciting designs these straight line spirals produce.

There are many variations of these spirals. Some of these variations have been connected to pursuit problems, such as the four spiders problem.

Another famous straight line spiral is the one produced by constructing square roots numbers using the Pythagorean theorem.

This spiral is formed from square root numbers using the Pythagorean theorem.

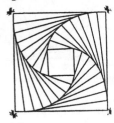

Four spiders start crawling from the four corners of a square. Each spider is crawling toward the one on its right, moving toward the center at a constant rate . Thus, the spiders are always located at four corners of a square. The curves formed by the spiders' path are equiangular spirals. The size of the initial square and speed of spiders determines how long it will take for the spiders to meet.

LOOKING AT THE DUALS OF PLATONIC SOLIDS

The Platonic solids— convex polyhedra with faces that are regular convex polygons— are only five in number. It is interesting to discover the connection between the Platonic and their duals.

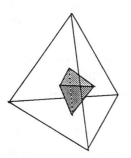

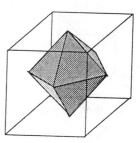

a tetrahedron & its dual

a cube & its dual

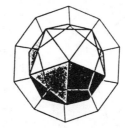

an octahedron & its dual   *a dodecahedron & its dual*   *an icosahedron & its dual*

To make the dual of a polyhedron, one first finds the center of each face and then joins the centers of faces that share a common edge. The illustrations reveal that the dual for a tetrahedron is a tetrahedron, the dual for a cube (hexahedron) is an octahedron, while an octahedron's dual is a cube. The dodecahedron's dual is an icosahedron, and an icosahedron's is a dodecahedron. None of the five Platonic solids' duals is anything but a Platonic solid!

DEFLATED POLYHEDRA — SCHLEGEL DIAGRAMS

tetrahedron

hexahedron (cube)

octahedron

dodecahedron

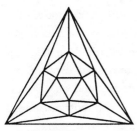

icosahedron

Each of these diagrams represents one of the five Platonic solids. They are called Schlegel diagrams, after the German mathematician Viktor Schlegel, who invented these particular diagrams in 1883. They remind one of 3-D objects which have been deflated. In essence, however, the polyhedron is inscribed in a sphere and projected through a point that is along the line of the north pole of the sphere. Its projection can change according to the polyhedron's position in the sphere. These Schlegel diagrams have the special property that one face of the polyhedron contains all the images of all the other vertices. In addition, the diagrams show all the polyhedron's vertices, edges and faces, even though their size and shape are distorted.

MATHEMATICS & THE ART OF M.C. ESCHER

The laws of mathematics are not merely human inventions or creations. They simply "are"; they exist quite independently of the human intellect. The most that any man with a keen intellect can do is to find out that they are there and to take cognizance of them. —**M.C. Escher**

And M.C. Escher certainly did take cognizance of mathematics. It is exciting to view many of his works through mathematical eyes.

Most of us are familiar with Escher's magical creations of the tessellations of a plane. His works far surpassed the traditional tessellations of a plane. He gives motion and life to the objects he tessellates, as illustrated in such famous works as *Metamorphosis, Sky and Water, Day and Night, Fish and Scales*, and *Encounter*. Besides transforming the plane, the tessellated objects undergo a transformation themselves. In addition, one sees his mastery of the concepts of translations, rotations, and reflections in periodic tiling.

Escher also used objects and ideas from the field of topology. The Möbius strip plays a key role in his woodcuts *Möbius Strip I* and *II* and *Horseman*. He exquisitely produces trefoil knots in his

A section from Escher's work *Metamorphosis III*.

work *Knots*. And even though Escher probably did not intend it, his *Snakes* is a perfect piece of art to introduce a topic on knot theory. *Print Gallery* and *Balcony* are wonderful examples of topolog-

*M.C. Esher's **Print Gallery** illustrates a topological distortion.*

ical distortion. These lithographs almost seem to be printed on rubber sheets which are magically distorted via topology.

Manipulation and mixing dimensions are other mathematical themes found in many of Escher's works. In *Reptiles*, Escher has 2-dimensional lizards eerily come to life in realistic crawling 3-dimensional forms. Similar transformations take place in *Magic Mirror* and *Cycle*. His use of concepts from projective geometry—

Reptiles *illustrates Escher's use of tessellations as well as his mastery of moving between the worlds of the 2nd and 3rd dimensions.* © 1994 M.C. Escher/Cordon Art-Baarn-Holland. *All rights reserved.*

perspective, traditional vanishing points, his own curved vanishing points—are responsible for the feeling of depth and dimension in *St. Peter's Rome*, *Tower of Babel* and *High and Low*.

Circles, ellipses, spirals, polyhedra and other solids are some of the geometric objects one finds in Escher's work. For example, *Three Spheres* creates a 3-dimensional illusion of spheres, although it is composed entirely of circles and ellipses. In *Stars* we find a variety of solids, including the Platonic solids, while a tetrahedron is the focus in *Tetrahedral Planetoid*. And in *Gravity* the stellated dodecahedron is present.

Escher makes the concepts of infinity come to life. No words are needed to define it, his work portrays its meaning. In *Whirpools* the spirals lead one's eyes on an endless journey. In *Square Limit* the feeling of infinite sequence approaching its boundary is projected. And *Circle Limit* would be an ideal model for Henri Poincaré's finite yet infinite non-Euclidean geometry. We get both the concepts of infinity and space tessellating in *Cubic Space Division*.

Lastly, in the field of optical illusion, Escher is phenomenal. His use of such impossible geometric figures as the Penrose tribar, help trick our eyes and confuse our minds. His *Waterfall* has us believing water is going upstream in an endless loop, while in *Ascending and Descending* there are two sets of people — one end-

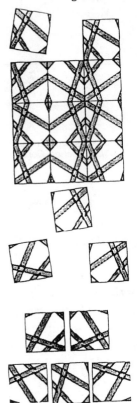

lessly climbing a staircase and the other descending a staircase all in a loop. Impossible figures are also his means for creating illusions in *Belvedere* and *Relativity*. In *Concave and Convex*, Escher is a master of oscillation illusion. Our eyes and mind are taken back and forth in the interior and exterior of an incredible structure and cast of characters. For example, one moment the vaulted dome is a roof and then within seconds it is a ceiling.

It is satisfying to explore Escher's work on many different levels. This introduction is but a glimpse of the wealth of mathematical ideas found in the work of M.C. Escher.

M.C. Escher mastered the art of tessellations. He also envisioned a tessellation stamp. The six faces of the cube have a design in six different positions. So as the stamp is rolled on a piece of paper, the design is tessellated. The design reproduced here was not done with a stamp. Instead, a computer was used to rotate and flip the design so that it could be tessellated.

TESSELLATING A PLANE WITH A MODIFIED RECTANGLE

Starting by modifying one of its lengths, translate that modification to the other length's position.

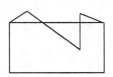

Now modify one of its widths, and translate the modification to the other width also.

Now take the resulting shape, and tessellate a plane.

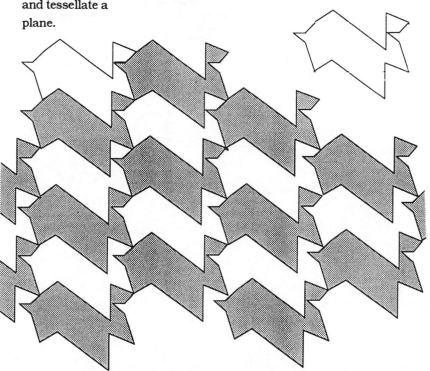

TESSELLATIONS OF OLD

From nature's tessellation of the honeycomb, to Roman mosaics, to the tiles of ancient Greece, to the marvelous designs by Moslem artists at The Alhambra, to M.C. Escher's phenomenal tessellations, to the simplicity of the Penrose tiles— tessellations cross centuries and cultures.

Roman mosaic floors have a variety of designs. This simple yet elegant pattern appears at the baths of Caracalla in Rome.

Here we have a tessellation design using pentagons and rhombi created by artist Albrecht Dürer. It dates back to the 15th century.

PROJECTIVE GEOMETRY & ART

Projective geometry is a field of mathematics that deals with properties and spatial relations of figures as they are projected—and therefore with problems of perspective. Many artists of the 15th century often worked in a number of different disciplines. Some of these artists, such as Giotto di Bondone (1266?-1337), Leon Battista Alberti (1404-1472), Piero della Francesca (1410?-1492), Leonardo da Vinci and Albrecht Dürer, studied engineering, mathematics, science and architecture, while pursuing their art. As a result Renainssance artists connected mathematical concepts, such as projective geometry, to the creation of their works. These artists developed techniques for painting realistic three dimensional scenes on two dimensional canvases. They analyzed how things changed when viewed from different distances and positions and developed ideas of perspective. They reasoned that if they perceived a scene outside through a window, it would be possible to project what they saw onto a window as a collection of points of vision if they kept their eye at a single focal point. The window would act as their canvas. The scene would naturally be affected by the position of the artist's eye and the position of the canvas. Some artists actually invented mechanical devices to aid drawing in perspective, as illustrated by the two diagrams.

Artists' works on per-

An Albrecht Dürer perspective instrument.

Leonardo da Vinci's spectograph.

spective influenced the development of projective geometry. Just as topology studies the properties of objects that remain unchanged after they have undergone a transformation, projective geometry studies properties of plane figures that do not change when they undergo projections. For example, when a circle is viewed from the side, it appears as an ellipse. Similarly a square would become a different shaped quadrilateral. Having been influenced by perspective in the art of

*This study of **The Adoration of the Magic** by Leonardo illustrates his use of linear perspective and a vanishing point.*

the Renaissance and desiring to help artists, 17th century architect/ engineer/ mathematician Girard Desargues initiated the study of projective geometry. He wrote a number of theorems (Desargues' Theorem) which are still essential in the study of projective geometry. His work influenced later mathematicians in this field, especially Blaise Pascal (1623-1662) and Jean Victor Poncelet (1788-1867). To create realistic three dimensional paintings Renaissance artists used concepts from projective geometry, namely— *point of projection, parallel converging lines,* and *vanishing points.*

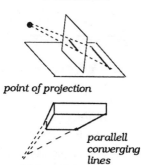

point of projection

parallell converging lines

vanishing point

This work by Jesuit monk Andrea del Pozzo (circa 1685), was painted on the hemi-cylindrical ceiling of St. Ignasio Church in Rome. His mural is an excellent example of perspective which illustrates the concepts of projective geometry of a single vanishing point. In fact, it may be considered to be too exact. A mark is located on the floor of the church, indicating where the viewer should stand to get the full effect the artist intended. At this location one actually feels the ceiling is infinite and the reality of *St. Ignatius Carried into Paradise.* Viewing the painting from any other point creates a distorted and uncomfortable effect.

MIXING ART & MATHEMATICS OF ALBRECHT DÜRER

...geometry is the foundation of all painting.
—Albrecht Dürer

Albrecht Dürer (1471-1528) was an artist of many talents. Of his 18 children, Dürer's father hoped it would be Albrecht who would follow him in his work as a goldsmith. As a result, Dürer was apprenticed to Michael Wohlgemut, who also happened to be an artist. After a three year apprentice, Dürer traveled and worked throughout Europe, where he learned painting, engraving, printing, and making woodcuts, all of which influenced his life's work. He felt the study of mathematics enhanced art, in particular geometry, perspective and ideas of projective geometry. He also did extensive work on the proportion of the human body. In addition,

A woodcut from Dürer's *Treatise on Geometry.*

he wrote books on the mathematics he used. Some of his work in mathematics was original and exact[1], while other work and geometric constructions were approximate and served his artistic needs. Along this line he developed mechanical devices to aid artists drawing in perspective.

The following sections illustrate how the work he pursued had a mathematical overtone, which helped him immeasurably in his art and other creations.

*In Dürer's woodcut **Melancolia** we find a magic square in the background, geometric solids and the sun's rays acting as lines of projection in this perspective drawing.*

---

[1]Dürer generated an epicycloid by following a fixed point of a circle as it rolled along the circumference of another circle, but lacking an algebraic foundation he did not analyze the curve. Likewise, he produced spirals from the projection of helical space curves, but did not pursue them mathematically.

CUBOCTAHEDRON-TRUNCUM

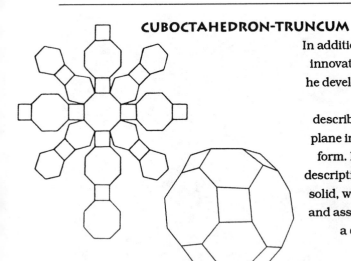

In addition to the many innovative techniques he developed, Dürer is credited with describing solids on a plane in unassembled form. Here is Dürer's description of one such solid, which, when cut and assembled, forms a cuboctahedron truncum.

DÜRER'S CONIC SECTION

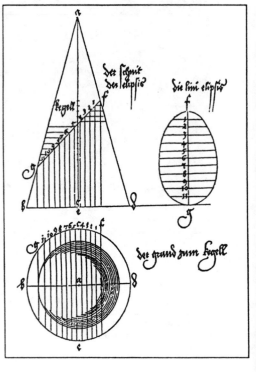

This illustration by Albrecht Dürer shows his interpretation of a conic section. His ellipse is somewhat egg-shaped, which implies he either believed the inclination of the plane cutting the cone made the ellipse narrower at the top or he made a slight error in calculation. Regardless, it is interesting to see his study of an ellipse.

From *Underqweysung der Messung mit dem Zircklelund Richtscheyt* (a treatise on geometric constructions) by Albrecht Dürer.

THE ORIGINAL PIXEL

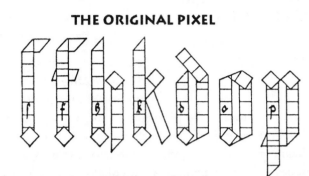

Dürer can be credited with launching the concept of topography. In one of his books he illustrates the geometric construction of Roman letters. Dürer devised his own mathematical methods for Gothic letters, by constructing each letter from a combination of squares, triangles, trapezoids and/or parallelograms. He systematized the structure of Gothic letters by using these little building blocks— somewhat similar to how letters are formed on a computer's monitor using the pixel[1] as the building block.

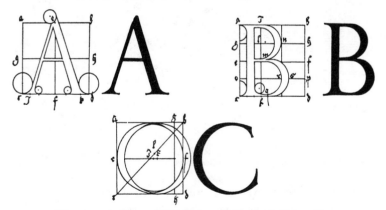

These geometric constructions were devised during the Renaissance, although the ideas were not published at that time.

---

[1]Pixel is the abbreviation for *picture element.* It is the visual representation of a *bit* on the computer's monitor. A *bit* stands for *binary digit,* which is the smallest unit of information that a computer can hold. The bit's value corresponds to the numbers 0 or 1, representing electrical flow as "off" or "on" which translates to a white or black square on the screen.

COMPUTER ART

Over the centuries the tools of an artist have ranged from a stick to a paint brush, to such devices as the camera obscura and the perspective window. Today artists are exploring a new form or medium of art that is linked to mathematics — the computer. Until recently computer art was produced by mathematicians, scientists, engineers, and just about everyone but artists. Initially, we were flooded with a great deal of art that resembled curve stitching, line work and optical illusions.

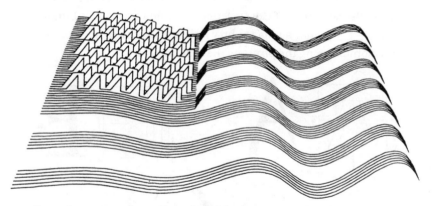

Computer generated art. Reproduced from **Computer Graphics** *by Melvin L. Prueitt. Courtesy of Dover Publications.*

Today, computers play a major role in commercial art. A computer skilled artist with advanced software can transform graphic art ideas for advertisements to multiple variations by changing type-styles, introducing various colors, scaling sizes of objects, rotating or flipping objects, duplicating various parts of an object in minutes. In the past, all of these changes would have taken the graphic artist hours if not days.

This oil painting effect was achieved by scanning a photograph and converting it via the computer.

Engineers, architects and other designers have not hesitated to embrace the computer in their creations. With just a few clicks of a mouse a building could easily be modified, a plane could be rotated to show all possible perspectives, cross-sections could be added, parts added or removed effortlessly. In the past this work was slow and painstaking.

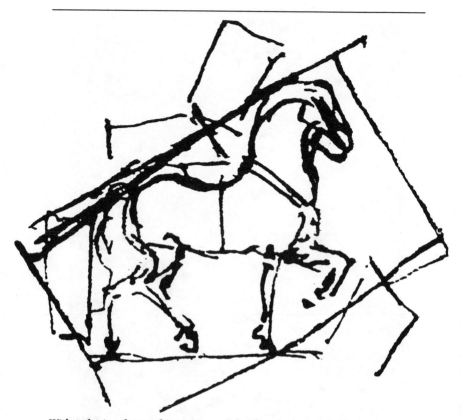

With today's advanced computers and software, Leonardo da Vinci might have rendered this sketch on a computer via an electronic sketch pad.

Artists over the centuries have used different mediums to create their works—watercolor, oils, acrylic, chalk, etc. There are some artists who feel the computer is an artificial means that lacks freedom of spontaneous expression.   They prefer the direct contact of their hands with the medium of their choice, rather than working by means of electricity on a keyboard and screen or with stylus and electronic tablet and screen.   Others view the computer as a challenge.

With improved software and hardware, colors can be mixed on the screen.   Rigid computer lines can be easily softened with any shaped curve. Changes can be made from an oil painting effect to a watercolor, and brush shapes can be changed at whim. Minute areas can be magnified and reworked more easily. Parts can be erased or cut out and pasted elsewhere in the work. The artist is definitely in charge of his or her creative work. The result can be either printed or entered on video, paper or film. Film leaves the option of enlarging it to any size. Perhaps a printer will be designed capable of capturing the texture of the work the artists create. Or perhaps we should consider this a new form of texture in itself.

Esteemed artists have displayed art created with a computer in well known international galleries, but it is still labeled *computer art*. Such works have yet to be simply labeled *art*.

What would Leonardo da Vinci think of the use of computers in art? Given his fondness for innovation[1], we can assume he would not have snubbed the use of the computer. He said *"...no human inquiry can be called science unless it pursues its path through mathematical exposition and demonstration."* His works reflect that he extended this to his art– for example, the predominant use of the golden rectangle in many of his works and the concepts of projective geometry in his masterpiece, *The Last Supper*.

One form of art should not be considered better than another, just different. Artists should be free to choose any means or medium.

---

[1]His notes and various innovations were used by artists to enhance and facilitate their artwork. Leonardo's mathematical inclinations led him to invent various types of special compasses capable of producing parabolas, ellipses and proportional figures. He is also credited with the invention of the perspectograph, used by artists to help draw objects in perspective.

$$1 = \frac{1}{2} + \frac{1}{2} = \sqrt{1}$$

$$= -e^{\pi i} = 1 \div 1 = \sum_{n=1}^{\infty} \left(\frac{1}{2}\right)^n$$

$$= 1 \times 1 = \text{googol}^0 = 1^{99}$$

$$= \frac{1000 - 1}{999} = \int_0^1 3x^2 dx$$

$$\sin(\pi/2) = -\left(\frac{X-Y}{Y-X}\right)$$

WITH X≠Y

THE MAGIC OF NUMBERS

The science of pure mathematics, in its modern developments, may claim to be the most original creation of the human spirit.—**A.N. Whitehead**

Some people think that the way certain numbers operate and the results that appear seem to possess a magical quality. Perhaps the illusion of hocus pocus is intensified because there are so many types of numbers — numbers which seem to be invented by mathematicians' whim.

In school we first learned about the whole numbers. Our first task was to learn to count, and then operate with them. Just when we began to feel that perhaps we had mastered the numbers having learned to add, subtract, multiply and divide, then seemingly out of the blue — fractions, decimals, integers,... — appeared. They

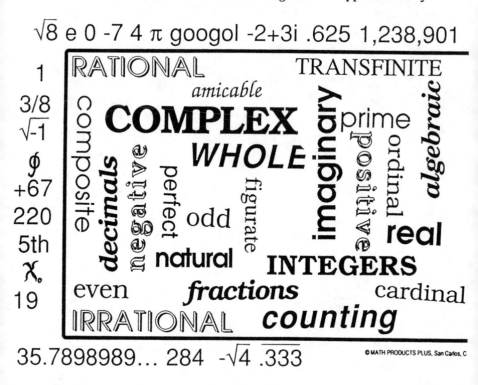

sometimes surfaced when particular problems didn't seem to have solutions. Then suddenly new numbers came to the rescue. For example, 7 divided by 2 was neither 3 nor 4, so we learned the number 3 and 1/2 did the trick.

What is so amazing is how many types of numbers there are, plus all the different classifications of numbers along with their individual characteristics. Numbers almost seem to be re-producing themselves. Here is a partial list of types of numbers with some descriptions— counting (also called naturals) numbers ({1,2,3,...}), whole numbers (include the counting plus the number 0), integers ({...-3,-2,-1,0,1,2,3...}), the rational (includes fractional, mixed numbers, improper fractions, proper fractions, decimals and repeating decimals), irrational (includes the never-ending non-repeating decimals, radicals without perfect roots, π, e, ϕ), real, imaginary, complex, transcendental, transfinite, perfect, ...

This section seeks to introduce you to some numbers, their properties and some of their quirky characteristics with which you may not be familiar, and to acquaint you more intimately with some you've known for a long time.

QUATERNIONS & THE GAMES NUMBERS PLAY

It is hard for the layperson to comprehend how can numbers be described as having various dimensions. It would seem that a number is just a number — something describing a particular quantity. How are such numbers as one, two, three, four, etc., dimensional? Well, leave it to the mathematicians to give another twist to the characteristics of numbers. For example, mathematicians consider any real number or any imaginary number as one-dimensional because they have one part to themselves that says their amount. Also they can be graphed on a line— a one dimensional object. On the other hand, the complex numbers are called two-dimensional numbers because they are composed of a real number and an imaginary number. For example when 5+2i is graphed, it is located on a plane (a 2-dimensional figure) called the complex number plane. Now you might ask how are 2-dimensional numbers used, and why does one need more than these. As with many mathematical ideas that are discovered or invented, their various uses come to light many years after they have been around. The complex numbers were no exception. Today, for example, they are used to describe patterns of flows in hydrodynamics, electric currents and the shapes of airfoils. Since these complex numbers were useful and so effective in solving various types of problems, it seemed that the next natural step was to look for three-dimensional numbers. Although the search for these was not successful, it lead to the discovery of four-dimensional numbers, the quaternions. These were invented by Sir William Hamilton in 1843. Like the complex numbers they were met with skepticism and suspicion.

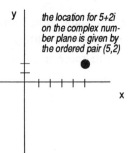

the location for 5+2i on the complex number plane is given by the ordered pair (5,2)

They are composed of four values or descriptive characteristics —
x, y, z-axes values (numbers that locate a point in
three-dimensional space) and a scalar (a quantity that does not
describe direction, as do x, y, and z values). The set of quaternion
numbers, unlike the set of real, imaginary or complex numbers,
are not commutative when it comes to multiplication — the order
in which they are multiplied makes a difference in the result. How
are these four-dimensional numbers being put to work today?
One of their everyday uses is communicating graphic information
on computers by giving descriptions of rotations in three dimensions. The following story tries to demonstrate that quaternions,
like all other numbers, have a personality of their own.

* * *

As usual at the numbers convention, the cliques were already
beginning to form. It was a shame that numbers could not think of
themselves as a happy family.

Over the centuries when just the counting numbers were around,
the odd and even numbers would argue about which was more
useful. But, they united forces when the integers entered the
picture with their negative numbers.

Now sides were already beginning to form over the big issue of the
convention — who would accept the newcomer, the quaternion?
The counting numbers had always been very elitist— only open to
whole numbers greater than or equal to 1. They were dressed in
their natural garb — all in increasing order with only one unit
between consecutive numbers. They had to study the newcomer,
and decide if it fell into its set. The integers were both hot and cold
about the quaternion, while zero, being neutral, never took sides,
since it was neither negative nor positive.

Surely the rationals would consider it more seriously. But the
fractions, as usual, were more interested in displaying their

numerators and denominators than to even talk to the decimals. Over the years, the decimals had gotten used to the fractions. They were no longer disturbed by the fractions' antics. The decimals knew how much easier they were to operate with than fractions, especially on calculators. .007 even called the fractions passé. But 1/7 jumped in to say "Although one has to find common denominators to add or subtract us, and we require some fancy foot work for multiplication and division, and while we prefer to be in lowest terms— some of your decimal representations of rational numbers are way out there. In fact, some calculator's memories can't hold your decimal representation." Thus the rationals, which included the counting numbers, the integers, fractions, and decimals, continued to fight among themselves.

Having seen the bickering that went on with the "rational" numbers, quaternion was understandably fearful of a group of radicals by the snack bar that included $\sqrt{2}$, $\sqrt{3}$, $\sqrt{15}$, and $\sqrt{6}$. Quaternion had heard how irrational they

could get. But to its surprise, they were interested in talking. "So I hear that you have a number of parts and are what they call 4-dimensional," $\sqrt{2}$ said. "Well, don't feel bad about it, I'm never ending non-repeating when in decimal form. All those digits are a drag to carry around so I prefer wearing my square root suit instead. Perhaps you too can find a more abbreviated way of expressing yourself."

Quaternion was a bit encouraged and felt more relaxed. Not wanting to get quaternion's hopes up; $\sqrt{3}$ added, "You'll need to be

cleared with the complex numbers set. It keeps track of all of us —
the counting, integers, rationals, irrationals, reals, and imaginary."

"But I heard the complex set has a split personality, and fluctuates
between the real and imaginary numbers," said quaternion.

Suddenly the complex number 3-5i walked over saying "You've got
that right, but the
complex number
plane gives
everyone of us
our own single
point on which to
reside. When
worse comes to
worst, I can
always take
refuge there. I
know it's my very
own point, no one
else has that
location, so there I can be alone, and regroup, relax and meditate.
We each have our own spot we can call home."

> Perhaps you too can find a more abbreviated way of expressing yourself.

$$\sqrt{2}$$

"You seem to have a multiple personality, what with your vectors
and scalar," 3-5i said to quaternion. " I'm sure the complex plane
has no place for you."

"I hope I can find my own point and home," quaternion said. With a
sad note to its voice it continued, "One doesn't know which way to
turn, or should I say which set to seek out."

"It certainly is difficult," said a rather deep voice. Quaternion turned
and saw π. "It was very hard for me to be accepted by the real
numbers. Although I'm irrational as $\sqrt{2}$ and the others, they would
not let me into the reals right away, saying there was no way to find

I heard the
complex set has
a split personality.

quaternion

my exact location on the real number line, unlike the $\sqrt{2}$, $\sqrt{3}$,
$\sqrt{5}$,... who use the Pythagorean theorem to find their locations. So
what was I to do? I had to do some fast talking, and the reals finally
realized how important an irrational number I am, especially since
all the circles depended on me for their circumference and area, and

I am
transcendental
to boot." "Well,
π, you speak as
though you're
the only
transcendental
number," said
the number e,
who was known
to be a braggart.
"I also happen
to be

The reals finally
realized how
important
an irrational
number I am.

π

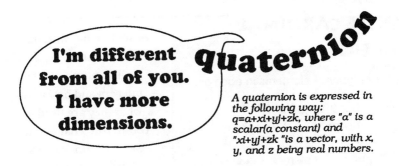

I'm different from all of you. I have more dimensions.

A quaternion is expressed in the following way: q=a+xi+yj+zk, where "a" is a scalar(a constant) and "xi+yj+zk "is a vector, with x, y, and z being real numbers.

transcendental, the base for the natural logarithms, and found extensively in nature besides being used in calculus."

Quaternion was beginning to get a headache and was tired of all the teasing and bickering. "Perhaps I don't belong here," quaternion said. "Perhaps you don't," all the complex numbers shouted. "But where do you belong?" they taunted quaternion.

"I'm different from all of you. I have more depth, more dimensions. Perhaps I belong to my own set. Yes, that's it! I am a member of the quaternions. The set of 4 dimensional numbers, since my general form is q=a+xi+yj+zk. " Having said this, quaternion began to rise from the convention floor, and suddenly disappeared, as if it had gone to another dimension.

CANTOR & THE INFINITE CARDINAL NUMBERS

$\aleph_0$ number of elements in each of the following sets —
{1,2,3,4,5,6,7,8,9,10,11,...}
{...,-4,-3,-2,-1,0,1,2,3,4,...}
{rational numbers}

$\aleph_1$ number of elements in each of the following sets —
{point on a line}
{point in a sphere}
{points in a cube}

$\aleph_2$ number of elements in each of the following set —
{all curves}

$\aleph_3$ number of elements in each of the following set —
set of ?

.
.
.

$\aleph_n$

We are familiar with the natural numbers, the whole numbers, the integers and how and what they are used to count. But what about Georg Cantor's transfinite numbers? How many are there and what type of things do they describe? Transfinite numbers describe the number of objects in a set. For example, A={apple, orange, pear} would use the cardinal number 3 to describe how many objects in set A. Between 1874-1895 Cantor studied and developed set theory. Since there are many sets of numbers that are infinite, it was apparent to him that a new set of cardinal numbers were needed to describe the cardinatlity of infinite sets. Hence, he created the transfinite numbers. $\aleph_0$, $\aleph_1$, $\aleph_2$, $\aleph_3$, $\aleph_4$, ... The symbol $\aleph$ stands for the Hebrew letter aleph. $\aleph_0$ (aleph-null) refers to the number of counting numbers. Any set that can be put into a one-to-one correspondence with the counting numbers is said to have $\aleph_0$ number of elements. The diagram above points out some sets with the cardinality of aleph-null $\aleph_0$, aleph-one $\aleph_1$, aleph-two $\aleph_2$, but no one has yet come up with examples of sets that use $\aleph_3$ or $\aleph_4$ or any higher ones.

NUMBER FANTASIES

The properties and workings of numbers at times seem almost magical. Choose any three digit number whose ones and hundreds digits are different.

For example, 285. Reverse the order of the digits. 582 Subtract the smaller from the larger number. 582-285=297 The result will always have a 9 as the tens digit, and the ones and hundreds digits always total 9. Now reverse the digits on the result. 792 Add these two numbers 792+297=1089 The result will always be 1089.

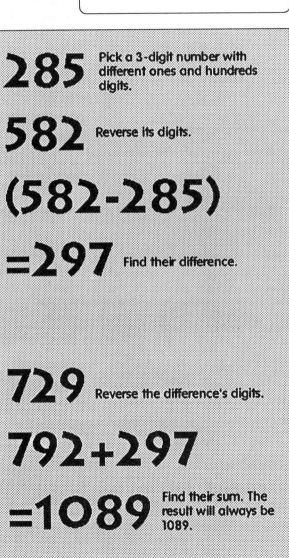

285 Pick a 3-digit number with different ones and hundreds digits.

582 Reverse its digits.

(582-285)

=297 Find their difference.

729 Reverse the difference's digits.

792+297

=1089 Find their sum. The result will always be 1089.

WHAT ABOUT PERFECT SQUARES?

A perfect square is a number which can be written as a whole number times itself. For example, 36 is 6x6 and 49 is 7x7.

Did you know?

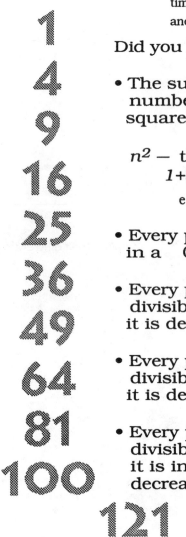

- The sum of the first *n* odd numbers is a perfect square,

  n^2 — that is
     $1+3+5+7...+(2n-1) =n^2$
     e.g. $1+3+5+7+9=25=5^2$

- Every perfect square ends in a   0, 1, 4, 5, 6, or 9.

- Every perfect square is divisible by 3 *or* is when it is decreased by 1.

- Every perfect square is divisible by 4 *or* is when it is decreased by 1.

- Every perfect square is divisible by 5 *or* is when it is increased or decreased by 1.

THE PARABLE OF π

Many, many years ago, there was a big party being given for the numbers of the time. One was there in all its glory. Two showed up with all the other even numbers in tow. And as many prime numbers as could be found had come. There were even some fractions like 1/2, 1/4 and 2/3. A few of the radicals had showed up like $\sqrt{2}$ *and* $\sqrt{7}$ *who had just arrived off the sides of a right triangle with 3. But*

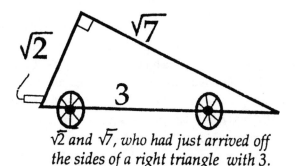

√2 and √7, who had just arrived off the sides of a right triangle with 3.

when π rolled in everyone asked, "Who invited you?". "What do you mean, 'Who invited me?', asked π. "I'm a number." "Yes you are, but do you know your location on the number line?" "What about $\sqrt{2}$*?" asked π. "Thanks, to Pythagoras and the use of a compass, I know exactly where I belong on the number line," replied* $\sqrt{2}$*.*

π felt embarrassed and hurt, but said, "I'm a little after the number 3."

"But exactly where?," they all chimed in.

Since 1 was a factor of every number, 1 felt π's pain and said, "Let's give π a chance to describe itself."

So π began to tell his story. "As you all know the Babylonians

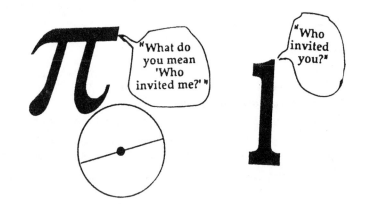

probably first discovered me. Some ancient scribe had drawn circles with different sized radii. The scribe took the diameter (by doubling the radius) of each circle. And just for kicks, decided to wrap each circle's diameter around it. To his surprise, he found that regardless of the size of the circle, its diameter always wrapped around it 3 and a little bit. This was an exciting discovery. The news spread quickly all over the world from Egypt to Greece to China. Everywhere people were learning about me. Because of my special connection to circles they were now devising new ways to find the area and distance around a circle by using me in their calculations. People were anxious to find my exact value. No offence, but they knew I was no ordinary number, especially since they had never come across a number quite like me. They were not able to derive me from any of their regular algebraic equations, so later on they also labeled me transcendental. You would have thought that people would have given up on finding my exact number name. I'm satisfied with π. It suits me just fine. But no, you know how stubborn some mathematicians are, they wanted to be more and more precise. So for the centuries that followed to the present, new means and methods have been developed to get a more accurate approximation.

The famous mathematician Archimedes found me to be between 3

10/71 and 3 1/7. In the Bible I appear twice and my value is given as 3. Egyptian mathematicians used 3.16 for me. And Ptolemy estimated me as 3.1416 in 150 A.D.

Mathematicians know they will never get my exact amount, but they keep on drawing me out to more and more decimal places. You can't imagine how much of a burden it is carrying around all those decimal places. Once calculus and computers are used, I'll be out to millions of places.

They say I am essential to computing various things, such as volumes, areas, circumferences, and anything that deals with circles, cylinders, cones, and spheres. I also play a role in probability. And with my millions decimal approximation, modern day

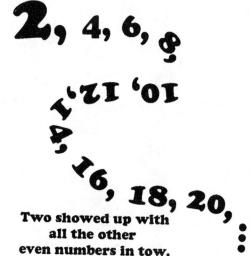

Two showed up with all the other even numbers in tow.

computers will rely on me to put them through their paces and test them for accuracy and speed.

"Say no more", shouted 1. 1 continued, " I'm sure we all agree that such a renowned number as π should be counted among us. After all, we know we each have our own point on the number line. No number can ever have another number's point. π has its point. It is not the most important thing about a number to know the exact location of its point."

"Agreed", shouted 3, one of the mystical numbers. " I think π lends the party a bit of mystery, variety and intrigue" said √2̄. "Welcome", chimed the rest of the numbers. "Let's get our party under way. Let's start counting", said π.

THE PRIME FASCINATION

We like to think
                    of ourselves
as the basic numbers.

2
                    3
5
                    7

We can describe
                    any whole number
                    uniquely.

11
                    13
17

Just break down   Just break down
the number   the number

any
                    whole number
to its prime factorization.

No two numbers
                    have the same set of
                    primes.

We're infinte in number, yet

Only one   2 is the
of us is even.   only even prime.

The rest of us
                    are odd.

The only factors
                    of any prime
are 1
                    and itself.

No one can   No one can
take us apart.   factor us further.
We're not composite.
We're prime!   We're prime!

One way to classify numbers is to describe them as either prime or composite. A prime number's only factors are 1 and itself. It is not divisible by any other number. On the other hand, a composite number has other factors beside 1 and itself (e.g. 12 is not prime because its factors are 1,2,3,4,6, and 12). In addition, every number can be described by a unique product of prime numbers (12's is $2 \cdot 2 \cdot 3$) — called its prime factorization. No other number besides 12 can be formed by multiplying two 2s and one 3. In the 1700s Christian Goldbach wrote to Leonhard Euler that he believed it could be shown that every even integer other than 2 is the sum of two prime numbers (e.g. 8=5+3; 28=13+15). This clear simple statement remains one of the unsolved problems of mathematics. Twin primes[1], Mersenne primes, Sophie Germain primes are among the other primes fascinations that mathematicians explore.

---

[1]Twin primes are primes that are only two units apart, such as 3 and 5. As of November 1993, the largest pair found was by Harvey Dubner $(1,692,923,232 \times 10^{4020} \pm 1)$ each with 4030 digits.

Reprinted from **Math Talk** by Theoni Pappas.
© 1991 Wide World Publishing/Tetra.

CANTOR & THE UNCOUNTABLE REAL NUMBERS

Georg Cantor's set theory and transfinite numbers were brilliant accomplishments. His proof of the countability of the rational numbers was elegant. Equally fascinating was his indirect proof showing that the real numbers are not countable. An infinite set is countable if its elements can be matched with the counting numbers. For the rational numbers[1], he devised a method for listing all possible fractions, and a method to match them up with the counting numbers. He tried the same type of method with the real numbers, but decided that they were not countable. So he set out to prove they were not countable by assuming they were, and therefore looked for a contradiction. First, he assumed he had an arrangement that listed all the real numbers. For example, for the numbers between 0 to 1 it might be a list like this—

.0 0 1 2 3 4 8 6 7 ...
.0 0 1 1 2 9 8 7 8 ...
.1 9 7 3 0 4 8 3 9 ...
.0 2 8 3 7 1 6 8 4 ...

Then he looked at the numbers along the diagonal of this list. He formed a number by taking the digits landing on the diagonal, and changed each digit on this diagonal to another digit (e.g. he may have changed them by simply adding 1 to each digit). Then by placing a decimal point in front of it, this number had to be between 0 and 1. Cantor argued that his list was ostensibly supposed to include all real numbers between 0 and 1, but did not include the one he just formed using the diagonal. This was the contradiction he was looking for. This number was supposed to be in the list, yet it differed from every term of the list at least at the digit that was along the list's diagonal. Thus, this diagonal number was a number that was between 0 and 1, and was not included in the "complete" list. Therefore, the list was not complete! This same proof could be used for any subset of the real numbers. Since the subset was not countable, the real numbers were not countable!

[1]Numbers that can be expressed as fractions or repeating decimals.

EUCLID'S PROOF OF UNENDING PRIMES

2, 3, 5, 7, 11, 13, 17, 19, 23, 29, 31, 37, 41, 43, 47, 53, 59, 61, 67, 71, ...

It seems the further out one ventures in the field of whole numbers, the primes become rarer and rarer. One might think because they appear less and less frequently, that perhaps they end somewhere. As early as about 300 B.C. Euclid provided the first proof that the prime numbers are infinite. He used indirect reasoning, in the following manner.

Suppose n represents the last prime number.

- *Now form a number from the product of all prime numbers up to and including the last one, n. 2x3x5x7x11x...xn.*

- *Add 1 to this product and call this number k.*
  k=2x3x5x7x11x...xn+1

- *k is either prime or composite.*

If k is prime, *then k is a prime number greater than the assumed largest prime, namely n, because k is 1 more than the product of the existing list of primes.*

If k were not prime, *then k must have a prime factor. k's prime factor cannot be one of the primes from the existing list of primes because we know none of these, namely 2, 3, 5, 7, 11, ..., n, can divide evenly into k — each of these primes would leave a remainder of 1. Therefore k must have a new prime number as its factor. Since there is no end to this process, the primes are infinite.*

As a matter of mathematics trivia— between 1 and 1000 there are 168 primes, between 1000 and 2000 there are 135, between 2000 and 3000 there are 127, between 3000 and 4000 there are 120.

NUMBER MAGIC

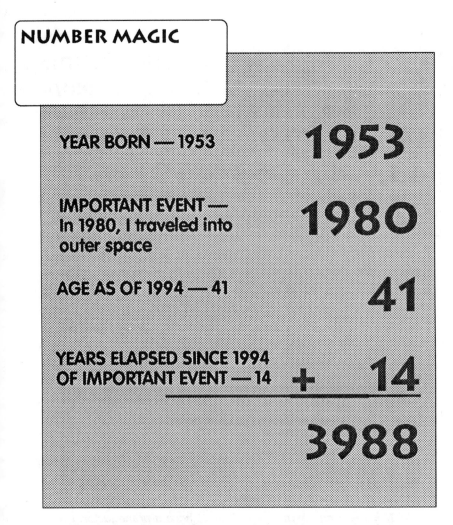

YEAR BORN — 1953   **1953**

IMPORTANT EVENT —
In 1980, I traveled into
outer space   **1980**

AGE AS OF 1994 — 41   **41**

YEARS ELAPSED SINCE 1994
OF IMPORTANT EVENT — 14   **+   14**

3988

Take the year you were born. To this add the year of an important
event in your life. To this sum add the age you will be at the end of
1994. Finally, add to this sum the number of years ago that the
important event took place. The answer is always 3988!

Every math enthusiast at one time or another has discovered or been delighted by tricks or oddities involving numbers. Here are two for you to explore and hopefully enjoy.

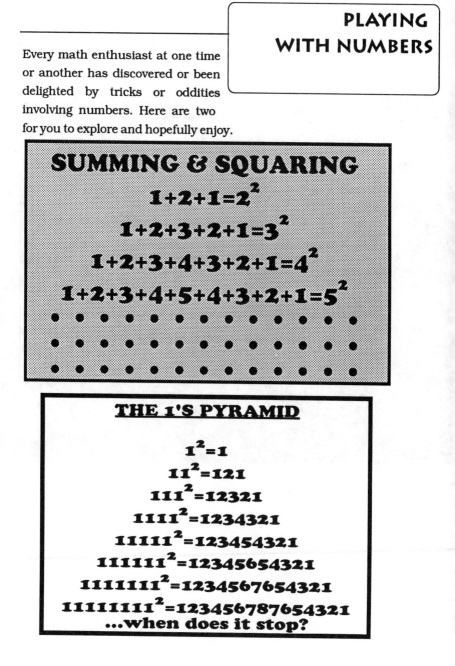

SUMMING & SQUARING

$$1+2+1=2^2$$
$$1+2+3+2+1=3^2$$
$$1+2+3+4+3+2+1=4^2$$
$$1+2+3+4+5+4+3+2+1=5^2$$

THE 1'S PYRAMID

$$1^2=1$$
$$11^2=121$$
$$111^2=12321$$
$$1111^2=1234321$$
$$11111^2=123454321$$
$$111111^2=12345654321$$
$$1111111^2=1234567654321$$
$$11111111^2=123456787654321$$
...when does it stop?

MATHEMATICAL MAGIC IN NATURE

WHAT HONEYBEES ARE BUZZING ABOUT
MATHEMATICS

HEXAGONS & NATURE

CHAOS IS FOR THE BIRDS

A CLOSER LOOK AT FRACTALS & NATURE

THE MATHEMATICALLY
ANNOTATED GARDEN

MATHEMATICS RIDES
THE CREST OF THE WAVE

*There is no branch of mathematics, however abstract,
which may not someday be applied to phenomena
of the real world. — **Lobachevsky***

*When we try to pick out anything by
itself we find it hitched to everything
else in the universe.*
—John Muir

Ever look at a leaf and wonder why it could be divided exactly in half, or notice the perfect shapes of stars in different flowers' petals, the spiral growth pattern of a certain shells, of pinecones, of the growth pattern of hair on a human head, or of the branches and bark of redwood trees? Nature abounds with examples of mathematical concepts. In our search to explain and understand how things are formed, we look for patterns and similarities that can be measured and categorized. This is why mathematics is used to explain natural phenomena.

As nature puts forth its wonders, most of us are oblivious to the massive calculations and mathematical work needed to explain something very routine to nature. For example, the Orb spider's web is a simple, but elegant natural creation. When this beautiful structure is mathematically analyzed, the ideas that appear in the web are indeed surprising — radii, chords, parallel segments, triangles, congruent corresponding angles, the logarithmic spiral, the catenary curve and the transcendental number e. Another example is the formation of plates on a tortoise shell. To explain the pattern formed, the mathematics of triple junction, hexagonal tiling, and calculus are needed. This chapter presents a few of nature's wonders and the mathematical ideas they conceal.

WHAT HONEYBEES ARE BUZZING ABOUT MATHEMATICS

Bees. . .by virtue of a certain geometrical forethought... know that the hexagon is greater than the square and the triangle, and will hold more honey for the same expenditure of material. — **Pappus of Alexandria**

Bees have not studied tessellation theory, nor orb spiders logarithmic spirals. Yet, as in so many things in nature, the architecture of insects and animals can often be analyzed mathematically. Nature uses the most efficient and effective forms — forms

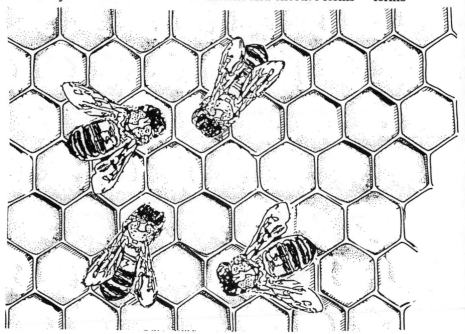

that require the least expenditure of energy and materials. Is it this that links nature and mathematics? Nature has mastered the art of solving maximum/minimum problems, linear algebra problems, and finding optimum solutions to problems involving constraints.

By focusing our attention on the honeybee, a wealth of mathematical ideas can be observed.

The square, the triangle and the hexagon are the only three self-tessellating regular polygons. Of the three, the hexagon has the smallest perimeter for a given area. This means that bees, when constructing hexagonal prism cells in the hive, use less wax and do less work to enclose the same space than if tessellating space with prisms of square or triangular bases. The comb's walls are made up of cells which are about 1/80 of an inch thick, yet can support 30 times their own weight. This explains why a honeycomb feels so heavy. A honeycomb of about 14.5"x8.8" can hold more than five pounds of honey, while it only requires about 1.5 ounces of wax to construct. The bees form the hexagonal prisms in three rhombic sections, and the walls of the cell meet at exactly 120° angles. The bees work simultaneously on different sections forming a comb with no visible seams. It is built vertically downward, and the bees use parts of their bodies as measuring instruments. In fact, their heads act as plummets.

Another fascinating "tool" a honeybee has is a "compass". Bees' orientation is influenced by the Earth's magnetic field. They can detect minute fluctuations in the Earth's magnetic field which are only discernible to sensitive magnetometers. This explains why the bees of a swarm occupying a new location simultaneously begin to build the hive in different parts of the new area without any bee directing them. All the bees orient their new comb in the same direction as their old hive.

In the illustration on the next page, the cells are shown to be closely packed, the bees having capped the ends with half rhombic dodecahedrons. In addition, the bees build the cell walls with a slope of 13°, which prevents honey from running out before the tops are capped with wax domes.

Communication is yet another area of interest. Worker bees returning to the hive from a scouting expedition communicate the direction of the food source they found by transmitting codes in the form of a "dance". They can communicate the direction

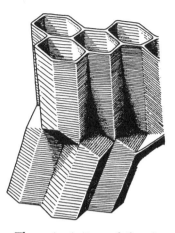

of the food and the distance. The orientation of the dance in relation to the sun gives the direction of the food, while the duration of the dance indicates the distance. It is equally surprising to learn that the honeybees "know" that *the shortest distance between two points is a straight line.* Perhaps this is the possible origin of the term "beeline". After a worker bee has collected a full load of nectar by randomly moving from flower to flower it knows to take the most direct route back to the hive.

The honeybee gets its mathematical training via its genetic codes. It is intriguing to analyze different aspects of nature from a mathematical point of view. This glimpse into the life of a honeybee is no exception. Here we find optimization of materials and work, tessellation of a plane and space, hexagons, hexagonal prisms, rhombic dodecahedron, geometric theorem, magnetic fields, codes, and amazing engineering techniques.

HEXAGONS

*The profound study of nature is
the most fertile source of
mathematical discoveries.*
—Joseph Fourier

Connections between mathematics and nature abound. Objects
and shapes from various fields of mathematics appear in many
natural phenomena.

*Nature also makes its hexagons in rock. Devil's Postpile National Monu-
ment, Mammouth Lakes, California. Photograph by Margaret Goodrich*

What is so special about the hexagon that nature repeatedly uses
it in many of its forms? The formation and growth of natural
objects is influenced by surrounding space and materials. The regular hexagon is one of three regular[1] polygons that tessellates a

Devil's Postpile National Monument, Mammouth Lakes, California.
Photograph by Margaret Goodrich

plane. Of the three (hexagon, square and equilateral triangle) the hexagon encompasses the largest area with the minimum amount of material. Another special feature of the regular hexagon is that it has 6 lines of symmetry, thereby allowing its shape to rotate numerous times without altering.   The sphere, which encloses the most volume using the least surface area is also connected to the hexagon. When spheres are placed side by side into a box (as illustrated at the bottom of page 126), each surrounded sphere is tangent to six other spheres. When line segments are drawn between these spheres at the tangent points, the shape circumscribing the sphere is a regular hexagon. Thinking of the spheres as soap bubble gives a simplified explanation of why a cluster of bubbles come together at what is known as the triple junction — three angles of 120°, the same degree measure of the interior angles of a regular hexagon. The triple junction appears in many areas, such as kernel formation on a cob of corn, the interior flesh of a banana, the cracking of dried earth. Discovering new occurrences of hexagons in nature is no less exciting than the first time they were observed on the back of a tortoise, in the beehive's honeycomb, or the shape of a crystal. Today scientists are equally fascinated with the sighting of hexagons in outer space. Since 1987 astronomers have been focusing

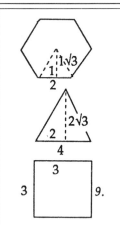

Suppose a perimeter of 12 units is available to form these three regular objects. The hexagon's area would come out to be $6\sqrt{3} \approx 10.4$. The triangle's area would be $4\sqrt{3} \approx 6.9$. The square's area is 9.

much attention on the Large Magellanic Cloud, where supernova 1987A was observed. It is not the first time gas bubbles have been seen following stellar explosions, but it is the first time the bubbles appear clustered in a honeycomb shape. Lifan Wang of the University of Manchester in England discovered the honeycomb, which measures approximately 30x90 light years, and is composed of about 20 bubbles whose diameters are about 10

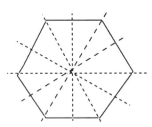

The six lines of symmetry of a regular hexagon.

light years. Wang suggests that a cluster, composed of similarly sized stars which have been evolving at about the same rate for several thou-

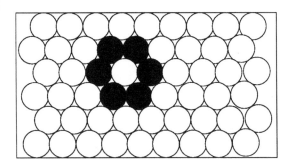

sand years, creates winds of such magnitude they shape the bubbles into the hexagonal configuration.

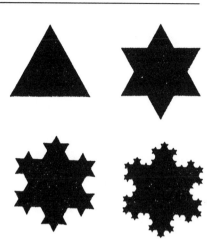

Lastly, a look at nature's snowflake illustrates hexagonal symmetry and fractal geometry. The snowflake possesses the shape of the hexagon. In addition, the growth of a snowflake is simulated by the Koch snowflake curve. This fractal is generated by an equilateral triangle as indicated in the drawings at the right.

These are the first four stages of the snowflake curve. Starting with an equilateral triangle, divide each side into thirds. Delete the middle third, and make a point of that length from the deleted part.

Consequently, the connection between equilateral triangles, the regular hexagon and the fractal snowflake links both Euclidean and non-Euclidean geometries.

Objects in nature have provided and do provide models for stimulating mathematical discoveries. Nature has a way of achieving an equilibrium and an exquisite balance in its creations. The key to understanding the workings of nature is with mathematics and the sciences. Galileo Galilei expressed it well when he said—*"the universe is written in ...the language of mathematics"*. Mathematical tools provide a means by which we try to understand, explain, and copy natural phenomena. One discovery leads to the next. What will the discovery of hexagons in outer space lead to? Only time will tell.

---

[1]A polygon is regular if its sides are the same length and angles the same size.

CHAOS IS FOR THE BIRDS

Ever become mesmerized by watching a flock of birds in flight as they swooped through the air, going from one direction to

A flock of seagulls. Aptos, California.

another in perfect harmony? Why don't the birds collide? Zoologist Frank H. Heppner wanted the answer to such a question. After painstakingly filming and studying the movie frames he had taken of flocks of birds, Heppner concluded that the birds were not guided by a leader. They flew in a state of dynamic equilibrium with the flocks leading edge continually changing birds at brief intervals. Until he was introduced to chaos theory and computers, he was unable to explain the flocking movement of birds. With concepts from chaos theory, Heppner

Perhaps a school of fish also moves in a state of dynamic equilibrium.

has now devised a computer program that simulates the possible movement of a flock of birds. He established 4 simple rules[1] based on avian behavior, and used triangles for birds. By varying the intensity of each rule, the flock of triangles would fly across the computer's monitor in familiar fashions. Heppner does not claim his program necessarily illustrates bird flock formation, but it does give a possible explanation of how and why flocks of birds move. It seems the chaos theory is at it again!

---

[1]The rules he established were: (1) Birds are attracted to a focal point or roost. (2) Birds are attracted to each other. (3) Birds want to maintain a fixed velocity. (4) Flight paths are altered by random occurrences such as gusts of wind.

A CLOSER LOOK AT FRACTALS & NATURE

Fractals have come to be referred to as the geometry of nature. Though there are abundant examples of objects from Euclidean geometry present in nature (hexagons, circles, cubes, tetrahedrons, squares, triangles...), the so often randomness of nature seems to produce objects which elude a Euclidean description. In these cases fractals provide the best descriptive tool. We know Euclidean geometry is great for describing such objects as crystals and bee hives, but one is hard pressed to find objects in Euclidean geometry to describe popcorn,

Lava flow from the Kilauea volcano, Hawaii

baked goods, bark on a tree, clouds, ginger root, and coastlines. While geometric fractals are used to describe such objects as a fern leaf or a snowflake, random fractals can be computer generated to describe lava flow and mountain's terrain. Euclidean geometry has its roots in ancient Greece (circa 300 B.C. with Euclid's work the *Elements*), while fractals have their origins in the late 1800's. In fact, the term fractal was not coined until 1975 by Benoit Mandelbrot. With their introduction, we have a geometry that can describe the ever changing universe.

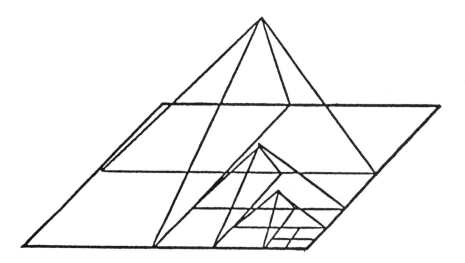

Fractals are considered fractional dimensions. In Euclidean geometry, a point is zero-dimensional, a line is 1-dimensional, and a plane is 2-dimensional. What about a jagged line? In the field of fractals, a jagged line's dimension is considered to lie between 1 and 2. Starting with a line segment and subdividing it into 3 sections as is done with a snowflake curve, its dimension will lie between 1 and 2. If we start with a rectangle, a 2-dimensional object, and subdivide it into 4 sections and construct a pyramid over its middle section, a fractal of dimension between 2 and 3 is formed.

FRACTALIZING THE SURFACE OF THE EARTH

Select a continent. Let's use South America as an example. Suppose we want to measure the distance of its coastline. In mathematics, we know that distance can be measured with any linear units we desire—for example, feet, inches, meters, miles, kilometers. Regardless of which we choose, they should represent the same distance—only expressed in different units. Yet if we measure the coastline of South America in kilometers and then in meters, our answers will differ if we consider the coastline to be fractal in nature. As our measuring units become smaller and smaller, we are able to measure more

and more inlets and bays of this coastline. In other words, theoretically we can get more detail as we continually decrease the size of our measuring unit which in turn continually increases the length of the coastline.

A coastline, in fractal geometry, is considered to be infinite in length because every little inlet and unit of sand is measured and the number of these units is constantly changing, similar to the formation of the snowflake curve curve (see page 127).

"Good morning day!" exclaimed the gardener, as she greeted the sunrise and her plants. Little did she know that strange things were lurking in the leaves and rich soil. Deep in the roots of the plants were *fractals* and **networks**, and from the cosmos, irises, marigolds, and daisies **Fibonacci numbers** were staring at her.

She proceeded about her daily ritual of tending to her garden. At each place, something unusual appeared, but she was oblivious, captivated only by the obvious wonders that nature presented.

She first went to clear out her ferns. Removing the dead fronds to expose the new fiddle heads, she did not recognize the **equiangular spirals** greeting her and the fractal-like formation of leaves on the ferns. Suddenly, as the breeze shifted, she was struck by the lovely fragrance of the honeysuckle. Looking over, she saw how it was taking over the fence and getting into the peas. She decided it definitely needed some judicious pruning. She did not realize that **helices** were at work, and the left-handed helices of the

THE MATHEMATICALLY ANNOTATED GARDEN

FRACTALS can appear as symmetrically changing/growing objects or as randomly asymmetrically changing objects. In either case, fractals are changing according to mathematical rules or patterns used to describe and dictate the growth of an initial object. Think of a geometric fractal as an endless generating pattern —the pattern continually

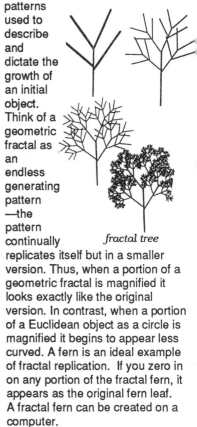

fractal tree

replicates itself but in a smaller version. Thus, when a portion of a geometric fractal is magnified it looks exactly like the original version. In contrast, when a portion of a Euclidean object as a circle is magnified it begins to appear less curved. A fern is an ideal example of fractal replication. If you zero in on any portion of the fractal fern, it appears as the original fern leaf. A fractal fern can be created on a computer.

NETWORKS are mathematical diagrams which present a simpler picture of a problem or situation. Networks were used by Euler in the Königsberg bridge problem (see *The Spell of Logic, Recreation, & Games* section). He reduced the problem to a simple diagram, which he analyzed and solved. Today networks are tools used in topology.

FIBONACCI NUMBERS 1, 1, 2, 3, 5, 8, 13, 21, ... Fiibonacci (Leonardo da Pisa) was one of the leading mathematicians of the Middle Ages. Although he made significant contributions to the fields of arithmetic, algebra and geometry, he is popular today for this sequence of numbers, which happened to be the solution to an obscure problem appearing in his book **Liber Abaci**. In the 19th century, French mathematician Edouard Lucas edited a recreational mathematics work that included the problem. It was at this time that Fibonacci's name was attached to the sequence. In nature the sequence appears in:

bloodroot

trilium

cosmos

wild rose

honeysuckle had wound around some of the right-handed helices of the peas. It required a careful hand to avoid damaging her new crop of peas.

Next she moved to weed beneath the palm tree she had planted to give her garden a somewhat exotic accent. Its branches were moving in the breeze, and she had no idea that **involute curves** *were brushing against her shoulders.*

She looked over at her corn smuggly. "Ha!" she thought. She had been hesitant to plant corn, but was encouraged by how well the young corn was progressing. Unbeknownst to her, **triple junctions** *of*

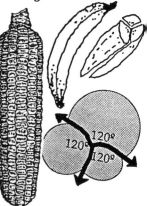

corn kernels would form within the ears.

How well the entire garden was shaping up and exploding with new growth! Admiring the new green leaves on the maple tree, she knew there was something inherently pleasing in their shape — nature's lines of **symmetry** had done their work well.

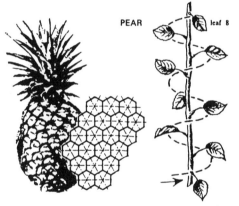

PEAR   leaf 8

And nature's **phyllotaxis** was only evident to the trained eye in budding leaves on branches and stems of plants.

Glancing around, she focused on the carrot patch. She was proud of how they were doing, and noted they need-ed thinning to insure uniform good sized carrots. She did not want to rely on nature to **tessellate** space with carrots.

•*Flowers with a Fibonacci number of petals* (trilium, wildrose, bloodroot, cosmos, columbine, lily blossom, iris)

•*Arrangement of leaves, twigs and stems* is known as **phyllotaxis.**

CHERRY   leaf 5

Select a leaf on a stem and count the number of leaves (assuming none have been broken-off) until you reach one directly in line with the one you selected. The total number of leaves (not counting the first one you selected) is usually a Fibonacci number in many plants, such as in elm, cherry or pear trees.

•*The pine cone numbers:* If the left and right handed spirals on a pine cone are counted, the two numbers are very often consecutive Fibonacci numbers. This also holds true for sunflower seedheads and seedheads of other flowers. The same is true of pineapples. Looking at the base of a pineapple count the number of left and right spirals composed of hexagonal shapes scales. They should be consecutive Fibonacci numbers.

SPIRALS & HELICES:

Spirals are mathematical forms which appear in many facets of nature, such as the curve of a fiddlehead fern, vines, shells, tornadoes, hurricanes, pine cones, the Milky Way, whirlpools. There are flat spirals, three dimensional spirals, right and left handed spirals, equiangular, logarithmic, hyperbolic, Archimedean spirals, and helices are just some of the many types of spirals which mathematics describes. The equiangular spiral appears in such growth forms of nature as the nautilus shell, a sunflower seedhead, the webs of Orb spiders. Some of the properties of the equiangular spiral are—angles formed from tangents to the spiral's radii are congruent (hence the term equiangular) — it increases at a geometric rate, thereby any radius is cut by the spiral into sections that form a geometric progression — its shape remains the same as it grows.

INVOLUTE CURVE: As a rope is wound or unwound around another curve (here a circle), it describe an involute curve. Involute is the shape found in the beak of an eagle, the dorsal fin of a shark, and the tip of a hanging palm leaf.

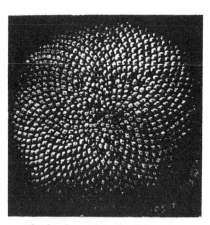

*She had no idea that the garden abounded with **equiangular spirals**. They were in the seedheads of the daisies and various flowers. Many things that grow form this spiral because of how it retains its shape while its size increases.*

It was getting warm, so she decided she would continue the cultivation when the sun shifted. Meanwhile, she made one final assessment — admiring the combination of flowers, vegetables and other plants she had so thoughtfully selected. But once more something escaped her. Her garden was full of spheres, cones, polyhedra and other geometric

shapes, and she did not recognize them.

As nature puts forth its wonders in the garden, most people are oblivious to the massive calculations and mathematical work that have become so routine in nature. Nature knows well how to work with restrictions of material and space, and produce the

Many types of symmetry appear in the garden. For example, in the above photograph one can find point symmetry in broccoli florettes and line symmetry in their leaves.

most harmonious forms. And so, during each day of spring, the gardener will enter her domain with a gleam in her eye. She will seek out the new growth and blossoms each day brings, unaware of the mathematical beauties flowering in her yard.

TRIPLE JUNCTION: A triple junction is the point where three line segments meet, and the angles at the intersection are each to 120°. Many natural occurrences result from restrictions caused by boundaries or availability of space. Triple junction is an equilibrium point toward which certain natural occurrences tend. Among other things, it is found in soap bubble clusters, the formation of kernels on the cob of corn, and the cracking of earth or stone.

SYMMETRY: Symmetry is that perfect balance one sees and senses in the body of a butterfly, in the shape of a leaf, in the form of the human body, in the perfection of a circle. From a mathematical point of view, an object is considered to possess line symmetry if one can find a line which divides it into two identical parts so that if it were possible to fold along that line both parts would match perfectly over one another. An object has point symmetry if infinitely many such lines exist for a particular point, for example a circle has point symmetry with respect to its center point.

TESSELLATE: To tessellate a plane simply means being able to cover the plane with flat tiles so that there are no gaps and no tiles overlap, such as with regular hexagons, squares, or other objects. Space is tessellated or filled by three-dimensional objects such as cubes, or truncated octahedra.

MATHEMATICS RIDES THE CREST OF THE WAVE

If you're a surfer, you know that sometimes it's difficult to know in advance when the surf will be up. Sometimes the surf appears perfect from the beach, but when you get into the water it's died down, so you wait for that perfect wave for what sometimes seems like hours. Other times the perfect waves come one after another and you have many to choose from. Needless to say wave theory and wave activity are a complex system. Many factors affect and create

Oceans waves. Aptos, California.

ocean waves. The wind, an earthquake, the wake of a boat, and of course the gravitational pull of the moon and the sun causing tides all disturb the ocean. Ocean waves travel on the surface of the water. These undulating forms have a somewhat random quality to them when there are multiple disturbances or factors interacting.

Much study was generated in the 1800's on the mathematics of ocean waves. Observations at sea and in controlled laboratory experiments helped scientists arrive at interesting conclusions. It started in Czechoslovakia in 1802, when Franz Gertner formulated the initial wave theory. In his observations he wrote how water particles in a wave move in circles. The water in the crest (highest point) of the wave moves in the direction of the wave and that in the trough (the lowest point of the wave) moves in the opposite direction. On the surface, each water particle moves in a circular orbit before returning to its original position. This circle was found to have a diameter equal to the height of the wave. There are circles being generated by the particles throughout the depth of the water. But the deeper the particle the smaller its circle. In fact, it was found that at a depth of 1/9 the wave length (the horizontal distance between two consecutive crests), the diameter of the circular orbit is about half that of the circular orbit of a surface particle.

Since waves are tied into these circling particles and since sinusoidal and cycloid shaped curves also depend on rotating circles, it is not surprising that these mathematical curves and their equations are used in the description of ocean waves. But it was discovered that ocean waves are not strictly sinusoidal or any other purely mathematical curve. The depth of the water, the intensity of the wind, the tides are only some of the variables that must be considered when describing waves. Today ocean waves are studied by applying the mathematics of probability, statistics, and complexity. A large number of small waves are considered and predictions are formulated from the data collected.

* * *

Some other interesting mathematical properties of ocean waves
are:

1) *The wave length depends upon the period.*

2) *The wave height does not depend on the period or the length
(there are some exceptions in which the influence of the
period and the length are minute).*

3) *The wave will break when the angle of the crest exceeds
120˚. When the wave breaks most of its energy is then
consumed.*

4) *Another way to determine when the waves will break is to
compare its height with its length. When this ratio is greater
than 1/7, the wave will break.*

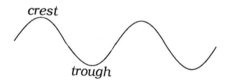

crest

trough

wave height   vertical distance from the crest to the trough

wave length   horizontal distance between two consecutive crests

wave period   the time in seconds it takes a wave crest to travel
one wave length

sinusoidal curve is a periodic (it
regularly repeats its shape) trigo-
nometric function .

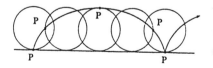

cycloid curve is the curve traced
by the path of a fixed point on a
circle which rolls smoothly on a
straight line.

Leonardo da Vinci's notebooks abound with sketches and writings on waves and water dynamics. This sketch from Codex Madrid II on folio 24 illustrates the swirling action of waves crashing on shore. Here he wrote, *"Nothing is carried away from the shore by the waves of the sea. The sea casts back to the shore all things left free out at sea. The surface of the water keeps the imprint of the waves for some time."* His descriptions of how waves break on shore and rebound onto the incoming waves were accurate conjectures that conformed to the the principles of wave motion as formulated in later years — namely that each point a wave strikes becomes the point of new disturbances and all disturbances affect the shape of the wave.

0	1	2	3	4	5	6	7	8	9	10	11
A	B	Γ	Δ	E	Ϛ	Z	H	Θ	I	I	IA
I	II	III	IV	V	VI	VII	VIII	IX	X		XI
0	1	10	11	100	101	110	111	1000	1001	1010	10

HINDU-ARABIC · BABYLONIAN · GREEK
EGYPTIAN HIEROGLYPHIC · CHINESE SCRIP
HEBREW · CHINESE ROD NUMERALS
ROMAN · EGYPTIAN HIERATIC
MAYAN · BINARY NUMERALS

MATHEMATICAL MAGIC FROM THE PAST

---

*In most sciences one generation tears down what another has built and what one has established another undoes. In mathematics alone each generation adds a new story to the old structure. — **Hermann Hankel***

In these days of conflicts between ancient and modern studies, there must surely be something to be said for a study which did not begin with Pythagoras and will not end with Einstein, but is the oldest and youngest of all.
*—**G.H. Hardy***

Discovering mathematical magic is not confined to present day things. History and ideas from the past hold a wealth of the magic of mathematics. One often wonders how the ancients made use of such ideas as irrational numbers, proofs, conic sections. If it were not for human curiosity and the desire to learn, would mathematics have progressed to where it is today? This chapter presents a few of the multitude of mathematical ideas that have surfaced over the centuries.

1
1 1
1 2 1 Pascal probability Laplace & Chu Shin–Chieh
1 3 3 1
Zeno paradox Whitehead • Einstein relativity
Liber Abaci Fibonacci sequence 1, 1, 2, 3, 5 ,8, . . .
al Khowarizmi algebra Bhaskara
$leg^2 + leg^2 = hypotenuse^2$ **Pythagoras** $a^2 + b^2 = c^2$
set theory **Cantor** transfinite numbers
Babbage computers Lovelace programming
Loyd puzzles *Dudeney* •Descartes Cartesian coordinates
Napier logarithms • calculator Pascal & Leibnitz
Apollonius conics Hypatia • Newton calculus Leibnitz & Seki Kowa
Agnesi curves & analysis **Noether** • Diophantus equations
Eratosthenes earth measurement
Euclid geometries Riemann & Bolyai & Lobachesky

Collage of mathematical ideas and their creators.

History shows that mathematical creativity is not privy to any particular culture, time period, civilization, or gender. The wealth of amazing ideas and contributions that were developed over the centuries is truly incredible and exciting to explore. Such an

NUMBERS SYSTEMS

0	1	2	3	4	5	6	7	8	9	10	11	12
ᛤ	Y	ᛒ	ᛗ	ᛟ	ᛝ	ᛞ	ᛠ	ᛡ	ᛣ	⟨	⟨Y	⟨ᛒ
A	B	Γ	Δ	E	Ϛ	Z	H	Θ	I	IA	IB	
I	II	III	IIII	II III	III III	III IIII	IIII IIII	IIII IIIII	∩	∩I	∩II	
ᐟ	=	Ξ	吅	五	大	七	八	九	亅	圡	圭	
א	ב	ג	ד	ה	ו	ז	ח	ט	י	אי	בי	
I	II	III	IIII	IIIII	T	ᴛᴛ	ᴛᴛᴛ	ᴛᴛᴛᴛ	⊟	⊣I	⊣II	
I	II	III	IV	V	VI	VII	VIII	IX	X	XI	XII	
⎮	⏝	⏝⏝	⎿ᴜⱼ	⅂	∠	⟨	⅂ᷧ	⥽	∧	∧I	∧II	
👁 •	••	•••	••••	—	•̲	••̲	•••̲	••••̲	=	≛	≛̲	
0	1	10	11	100	101	110	111	1000	1001	1010	1011	1100

HINDU-ARABIC·BABYLONIAN·GREEK·EGYPTIAN HIEROGLYPHIC·CHINESE SCRIPT·HEBREW
CHINESE ROD NUMERALS·ROMAN·EGYPTIAN HIERATIC·MAYAN·BINARY NUMERALS

exploration would take the reader on a trip that traverses time and countries all over the world. Such a voyage would reveal that some ideas were discovered almost simultaneously in different countries, as was the case with the development of hyperbolic geometries. We know that numbers and numeration systems of some form are indigenous to all peoples. We find that the use of place value and the number zero was developed in many parts of the world — first by the Babylonians with their base 60, later the Maya with a modified base 20 system and by the Hindu when they developed and evolved a positional notation for the base 10 system. We find that their system was later improved upon and standardized by the Arabs. The Chinese also developed a place value system and used zero with their *rod numerals*, which they evolved into a sophisticated base ten numeration system used especially in performing calculations. Only a few of the many

FERMAT	HYPATIA	POINCARÉ	PYTHAGORAS
EULER	NOETHER	BHASKARA	OMAR KHAYYAM
PLATO	I-HSING	RIEMANN	LOBACHEVSKY
PASCAL	LAGRANGE	LEIBNIZ	ARCHIMEDES
GALOIS	CAUCHY	HILBERT	RAMANUJAN
EUCLID	EUXODUS	DEDEKIND	FIBONACCI
CAYLEY	JACOBI	DESCARTES	KOVALEVSKAYA
PAPPUS	SOCRATES	BOOLE	KRONECKER
THALES	HAMILTON	SEKI KŌWA	CAVALIERI
CANTOR	SACCHERI	KUMMA	BERNOULLI
BOLYAI	FOURIER	PTOLEMY	APOLLONIUS
AGNESI	GALILEO	GAUSS	WEIERSTRASS
GÖDEL	HERMITE	LEGENDRE	De MOIVRE
NAPIER	LAPLACE	SYLVESTER	ERATOSTHENES
ZENO	NEWTON	LOVELACE	KEPLER
ABEL	GERMAIN	DIOPHANTUS	ARISTOTLE
RUSSELL	EINSTEIN	SOMERVILLE	AL-KHOWARIZMI
LUI HUI	BIRKHOFF	du CHATELET	WHITEHEAD

An array of some famous mathematicians from the past.

thousands of mathematicians and ideas are mentioned in this chapter. Those I discuss in no way reflect their superiority or stature over those not appearing. **I urge you to use these sections as spring boards for further study and to enhance your understanding of the magic of the past.** This is the spirit in which illustrations of the array of famous mathematicians, of the collage of mathematicians and their ideas, and of the diagram of numbers systems is presented. We are in the midst of many *new* mathematical discoveries. One does not have to be a mathematician to understand the essence of these ideas or appreciate their creativity. Seek out the magic in the past and the present.

BABYLONIANS & SQUARE ROOTS
WHAT WERE THEY DOING WITH ACCURATE SQUARE ROOT APPROXIMATIONS?

We often think of ancient mathematics as just that — ancient and remote! Yet looking back, it is surprising to discover you may be using a similar idea, value, or concept that people used thousands of years ago. What we have learned about Babylonian mathematics comes mainly from a few clay tablets of cuneiform

text from archeological excavations. The tablets date around 3000 to 200 B.C.[1]. They reveal that the Babylonians dealt with the following mathematical concepts—

- *equations in one unknown*
- *systems of equations in two unknown, tables of approximations*[2]

- *volumes and areas*
- *calculation of the area of a triangle and a trapezoid*
- *the approximation of π, 3 was used in determining the area of the circle— 3r²*
- *the volume of prisms and cylinders by multiplying area of the base times it height*
- *the Pythagorean theorem*
- *aspects of number theory, for example :*

$$1+2+4+...2^9=2^9+(2^9-1)$$

The Babylonian tablet[3] on the previous page illustrates an astonishingly accurate approximation for $\sqrt{2}$. It is equally amazing to learn that the Babylonian sexagesimal positional number system

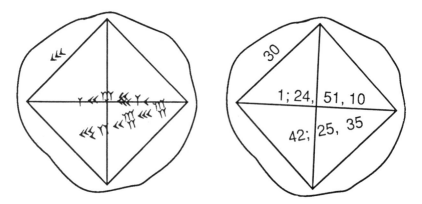

lent itself to such accuracies. The Babylonians updated the Sumerian number system to a sexagesimal positional number system.[4] It was the first positional number system of its time. Initially lacking a zero and a sexagesimal point, it relied on the context to indicate the intended value of the number.   For example, this number

could represent

$$11(60) + 12 = 672 \text{ or } 11+ 12/60.$$

The Babylonians later developed or for a zero place holder.

What were the Babylonians doing with such accurate approximations for $\sqrt{2}$? Examining this cuneiform tablet closely we see that the figure is a square with diagonals drawn. One side of the square has the symbol which is how they wrote 30.

Along the horizontal diagonal they wrote

which represents 1, 24, 51, 10. By assuming the sexagesimal point is between 1 and 24, the number converts to
$1 +(24/60) +(51/60^2) +(10/60^3)$
$$= 1+(2/5)+(51/3600)+(1/216000) \approx 1.4142129+$$
which compares to the $\sqrt{2}=1.414213562....$ To arrive at their estimate the Babylonians probably used a repetitive approximation method often used by the Greeks.[5]

We know the Babylonians had a good understanding of the Pythagorean theorem. The value they calculated for the vertical

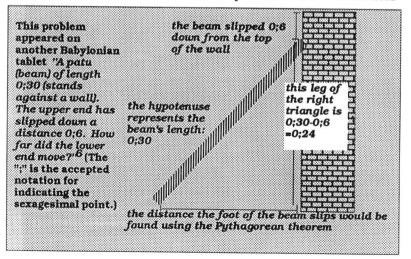

This problem appeared on another Babylonian tablet "A patu (beam) of length 0;30 (stands against a wall). The upper end has slipped down a distance 0;6. How far did the lower end move?"[6] (The ";" is the accepted notation for indicating the sexagesimal point.)

the beam slipped 0;6 down from the top of the wall

the hypotenuse represents the beam's length; 0;30

this leg of the right triangle is 0;30-0;6 =0;24

the distance the foot of the beam slips would be found using the Pythagorean theorem diagonal is an accurate approximation for the length of the square's diagonal. Namely, the diagonal numbers 42; 25, 35 convert to $42+(25/60)+(35/3600) \approx 42.42638889$ while $30\sqrt{2} \approx 42.42640687$. In addition to working with right triangles whose sides were rational numbers (e.g. {3,4,5}, {5,12,13}), they also used the Pythagorean theorem on right triangles whose sides were not all rational numbers. This explains their use of approximations for such irrational numbers as $\sqrt{2}$.

---

[1]Translations of O. Neugebauer, *Mathematics Keilschriftexts*, 1935-1937. F. Thureau—*Dangin Textes Mathematiques Babyloniens* 1938, and O. Neugebauer and A. Sachs *Mathematical Cuneiform Texts* . To expedite numerical translation of the tablets, O. Neugebauer developed the notation of commas to separate place values and " ; " to represent the sexagesimal point.

For example, ◀Υ Υ◀ΥΥΥ translates to 11, 1, ; 10, 3 which

means $11(60) +1 +(10/60) + (3/3600)$.

[2]Tablets have revealed tables for approximating various mathematical values. For example: tables of values for reciprocals which were used when division had a remainder, and tables of values for square roots and cube roots have been excavated.

[3]This is from tablet YBC7289 of the Babylonian collection at Yale.

[4]The Babylonians used two symbols to write their numbers. Υ for one and ◀ for ten. The position of these symbols determined their value.

[5]The ancient Greeks used the following method to approximate square roots: assume their first guess for 2 was a (for example, let a=1). Their next estimate was determined by taking 2/a (which equals 2/1=2). Then their next estimate was the average of these two estimates, (1+2)/2, which equals 1.5 . The estimate 1.5 would then be averaged with a new approximation, 2/1.5. This process would be continued.

[6]From *Science Awaking* by B.L.van der Waerden., Jon Wiley & Sons, Inc., New York, 1963.

THE LADDER THAT INCHES UP ON THE √2

The ancient Greeks discovered how to construct the lengths of irrational numbers using the Pythagorean theorem. They used inscribing and circumscribing regular polygons and the concepts of infinities and limits to close in on the area of a circle. They also developed a type of ladder arithmetic using ratios to approximate the value of irrational numbers. Here's how it worked for $\sqrt{2}$.

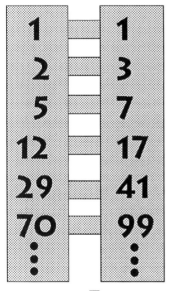

Starting with the 1 and 1 from the top of the latter, the rest of the numbers in the first column are generated in the following way

Here is how the numbers from the first column generate the the numbers in the second column

1+1=2	1+2=3
2+3=5	2+5=7
5+7=12	5+12=17
12+17=29	12+29=41
29+41=70	29+70=99

The ratio $1{:}\sqrt{2}$ is squeezed in on by the ratios of the numbers on the same rung of the ladder. The ratios continually come closer and closer to $1/\sqrt{2}$. The limit of ratios is the value of $1/\sqrt{2}$.

NOTE: The two numbers at each rung of the ladder solve the equation—
$y^2 - 2x^2 = \pm 1$. The x values are the numbers on the left side of the ladder.

$1/\sqrt{2}$ =.707106781...
$1/1 = 1$
$2/3 = .666...$
$5/7 = 71428571429...$
$12/17 = .70588235294...$
$29/41 = .70731707317...$
$70/99 = .7070...$

THE CHINESE METHOD OF PILING SQUARES

Finding experts who are able to translate ancient Chinese writing is difficult. Finding experts able to translate manuscripts that deal with mathematical ideas is even more difficult. This explains why Chinese examples of mathematical themes are scarce. *Hsuan-thu*, the piling of squares, was a

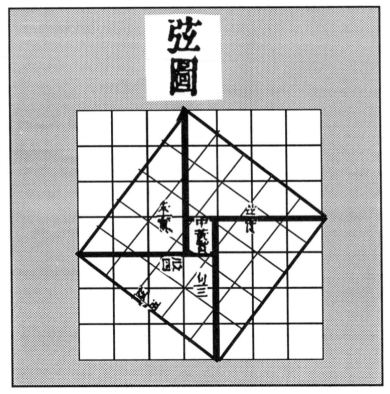

technique that Chinese mathematicians used in order to arrive at algebraic conclusions using geometric and arithmetic means. This particular illustration is from the Chinese manuscript, *Chou Pei*. The date of *Chou Pei* is disputed, with possible dates ranging from

1200 B.C. to 100 A.D. . If 1200 B.C. is accurate, it would be one of the earliest known demonstrations of the Pythagorean theorem, predating Pythagoras and the Pythagoreans.   The Pythagorean theorem has appeared in many civilizations thoughout history. In architecture, it was one means of assuring the formation of a right angle. In mathematics is has been and is an indespensible tool crossing many mathematical disciplines.

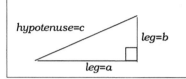

The Pythagorean theorem states that for any right triangle the sum of the squares of its two legs equals the square of its hypotenuse ($a^2 + b^2 = c^2$). (The converse is also true.)

In the left diagram below, the interior square's area is indicated as 5x5 or 5^2 = 25 square units. It has been subdivided into 4 right triangles, each of area (1/2)(3x4) and a square of area 1x1, totaling 25 square units. In the right diagram below, the square is divided into two smaller overlapping squares, one 3x3 and the other 4x4. The part they overlap has the same area as the vacant part of the 5x5 square they do not occupy, which illustrates that the area of larger square (5^2) equals the sum of the two smaller squares' area, namely 3^2 and 4^2.

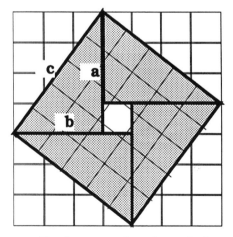

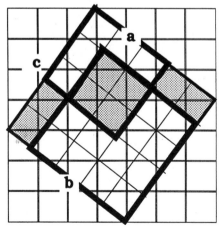

This diagram explains how to find the area of the interior shaded square by summing the areas of 4 triangles and the unit square in the middle. In general it shows—

$$c^2 = 4(1/2)ab + (a\text{-}b)^2$$
$$= 2ab + (a^2\text{-}2ab\text{+}b)^2$$
$$= a^2 + b^2$$

The sum of the two shaded rectangles' areas equals the area of the small shaded square (this is the square created by two overlapping squares). Letting 5, 4, and 3 take on the variable c, a and b, it shows

$$a^2 + b^2 = c^2.$$

ONE OF THE EARLIEST RANDOM NUMBER GENERATORS

Although this die was not referred to as a random number generator during ancient Greece, it still gets credit for being one of the earliest remaining die. Today it appears in the National Archeological Museum in Athens, Greece. Dice have played many roles over the centuries. They have been used to foretell the future, to implement moves of various games such as

backgammon and monopoly, or, as in craps, they are the main elements of the game. Mathematicians have long been intrigued with dice from the viewpoint of probability. In fact, dice can be considered responsible for getting Blaise Pascal and Pierre de Fermat to focus their attention on probability. While gambling, Pascal was asked by a friend how the pot should be split if the game were to be stopped before it was over. Pascal wrote Fermat about the problem. In 1654 the two men worked out their theory of probability in their correspondences, and thus launched this new branch of mathematics. Today dice and other shaped random number generators are used to teach various aspects of the theory of probability.

EGYPTIAN MULTIPLICATION

The Egyptian method of multiplication survived and spread for centuries and over civilizations. In ancient Greek schools it was taught as *Egyptian calculation*. In the Middle Ages its techniques were taught and referred to by specific names such as *duplatio* for doubling and *mediatio* for halving. Here is an example, from the Rhind papyrus of how an

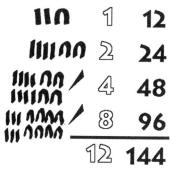

Egyptian scribe would have multiplied 12x12. One first starts with 12. Then it is doubled to give 24, which is then doubled to give 48, which in turn is doubled giving 96. Slashes are drawn next to the 4 and the 8, indicating their sum is twelve. Then their corresponding amounts are added, giving the answer, 144. The Egyptian method of multiplication eliminated the need for memorizing multiplication tables, since it relied mainly on addition.

Division was done similarly. To divide 1120 by 80, you find what you would multiply 80 by to get 1120. Depending on how large a number you are dividing, the divisor is either doubled, or multiplied by 10, 100, 1000, etc. The results can then be doubled until a sum is found that equals 1120. If the problem did not come out even, the Egyptians used fractions. As in the example 47÷33.

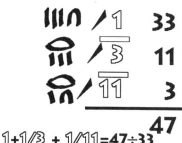

$1 + 1/3 + 1/11 = 47 \div 33$

THE FIRST SCIENTIFIC LABORATORY

Ancient Greek mathematicians and scholars are often considered to be predominately theorists and philosophers. Now new research[1] has revealed that the first recorded laboratory for experimentation was set up by Pythagoras or the Pythagoreans (6th century B.C.).

It is not surprising that there are no records from the Pythagoreans to support this claim, since the Pythagoreans were sworn to secrecy. There is supportive evidence from later scholars. This illustration is from the writings of the Roman scholar Boethius (5th century A.D.). It shows Pythagoras experimenting with sound, specifically the relationship between an object's proportions (in this case bells) and the tones they

produce. Theon of Smyrna (2nd century A.D.), along with other ancient authors, wrote of similar types of experimentations by other Greeks as well as the Pythagoreans. If one considers the inventions of the Greek mathematician Archimedes (287-212 B.C.)— laws of lever and pulleys, methods for comparing volumes of objects by submergence in water, the Archimedean screw, the catapult — one realizes the existence and importance of experimentation.

---

[1]These new findings were presented by Andrew D. Dimarogonas of Washington University in St. Louis in the *Journal of Sound & Vibration,* May, 1990.

PLATO DOUBLES THE SQUARE

ΑΓΕΩΜΕΤΡΗΤΟΣ ΜΗΔΕΙΣ
ΕΙΣΙΤΩ

Let no one ignorant of geometry enter here.

These words were over the doors at the Academy of Plato in Athens. Although Plato (428-348 B.C.) is not famous for his mathematical contributions, he is renowned for providing a center where he guided, encouraged and inspired mathematical thought. This elegant method for doubling the area of a square appears in Plato's dialogue *Meno*. The diagram illustrates how to double a square and how not to double a square.

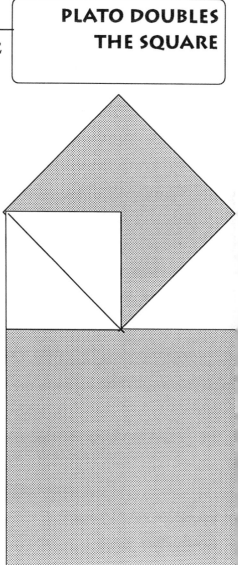

The upper shaded square's area is double that of the white square. The lower shaded square has area four times that of the white square. Note that although its side is twice the size of the white square's side, its area is quadrupled.

HOW THE ROMANS FIGURED THE AREA OF A CIRCLE

To find the area of any given circle, the Romans used the area of a square whose diagonal was 1/4 longer than the circle's diameter.

Here's how it worked, and its accuracy.

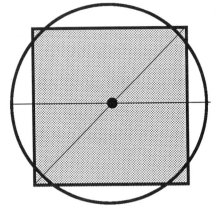

Suppose this circle's diameter is d.

Then this makes the square's diagonal d+.25d = 1.25d.

Using the Pythagorean theorem, the square's side is $1.25d/\sqrt{2}$.

Squaring the side of the square gives the square's area as $1.5625d^2/2$.

Since the circle's radius is .5d, and the area formula for any circle is πr^2, the actual area of the circle is $(.5d)^2\pi=.25d^2\pi$.

Using 3.1416 for the approximation of π. We get —

ROMAN METHOD   vs   ACTUAL AREA METHOD

$$1.5625d^2/2\approx.78125d^2 \qquad .25d^2\pi \approx.785d^2$$

Equating **.78125d²** & **.25d²**π *and solving this relation for π we get what approximation the Romans had for π, namely—*

$$.78125d^2 = .25d^2\pi$$

$$\frac{.78125d^2}{.25d^2} = \pi$$

$$3.125 = \pi$$

$$3+1/8 = \pi$$

HOW THE GNOMON TRISECTS AN ANGLE

Trisecting an angle was one of the three famous impossible construction problems of antiquity which launched many mathematical discoveries. Although an angle cannot be trisected using only a compass and straight edge, it can be trisected using a tool the Greeks referred to as the gnomon. The gnomon can be used to make and determine right angles. It was used by the ancient Greeks to trisect an angle in the following way:

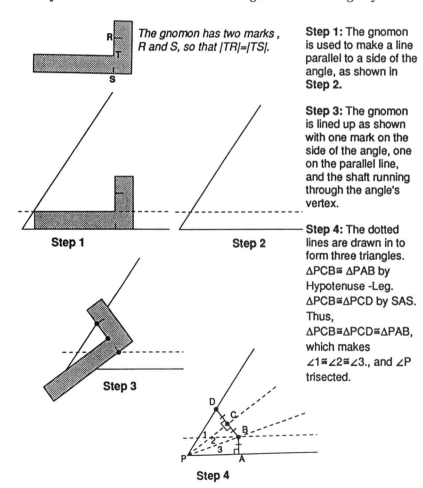

The gnomon has two marks, R and S, so that |TR|=|TS|.

Step 1

Step 2

Step 3

Step 4

Step 1: The gnomon is used to make a line parallel to a side of the angle, as shown in **Step 2.**

Step 3: The gnomon is lined up as shown with one mark on the side of the angle, one on the parallel line, and the shaft running through the angle's vertex.

Step 4: The dotted lines are drawn in to form three triangles. ΔPCB≅ ΔPAB by Hypotenuse -Leg. ΔPCB≅ΔPCD by SAS. Thus, ΔPCB≅ΔPCD≅ΔPAB, which makes ∠1≅∠2≅∠3., and ∠P trisected.

UNSOLVED MATHEMATICAL MYSTERIES

Mathematics certainly offers an abundance of problems. In fact, mathematics and problems are inseparable. History shows that mathematical ideas have been catalysts for mathematics problems, and that math problems have stimulated many mathematical ideas and discoveries. The *three impossible construction problems of antiquity*[1]; *the Königsberg Bridge Problem*[2] and *the Parallel Postulate Problem*[3] are historic examples of problems that have been resolved and in the process stimulated mathematical thought, ideas and discoveries. The posing and pondering of mathematical problems and questions and the scrutinizing of solutions and proofs are driving forces for mathematicians.

Here are a few famous "unsolved" mathematics problems:

THE UNSOLVED PRIME NUMBER PROBLEMS_____

• Is there one formula or a test to determine whether or not a given number is prime?

• Is there an infinite number of prime pairs? *A prime pair is a pair of consecutive primes whose difference is two.* For example, 3 and 5, since 5-3=2. Some others are 5 & 7, 11 & 13, 41 & 43.

• *The odd perfect number mystery.* A perfect number is one which is equal to the sum of its proper divisors (a proper divisor is a divisor other than the number itself). The number 6 is an example of an even perfect number because 6=1+2+3. Other examples are 28, 496, and 8128. Circa 300 B.C., Euclid showed that if 2^{n-1} was a prime number that $2^{n-1}(2^n-1)$ is a perfect number. Then, in the 18th century, Leonhard Euler proved that *any even perfect number* must comply with Euclid's form.
For example, $8128=2^6(2^7-1)$.

But odd perfect numbers remain a mystery. As yet no one has

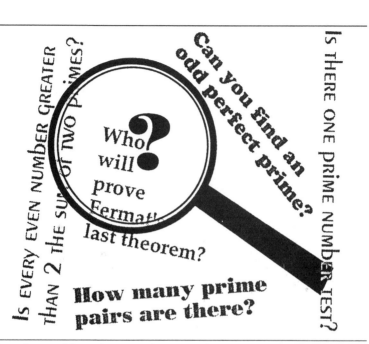

found an odd perfect number, nor has anyone proved that all perfect numbers are even.

GOLDBACH'S CONJECTURE

Is every even number greater than 2 the sum of two primes?

In 1742 German mathematician Christian Goldbach (1690-1764) wrote to Leonhard Euler (1707-1783) the conjecture *that every even number except 2 was the sum of two primes.* Examples: 4=2+2; 6=3+3; 8=3+5; 10=5+5; 12=7+5, ... Although Goldbach's conjecture is believed to be true, no one has proven it so as yet. Thus far the following ideas have been shown: In 1931, Soviet mathematician L. Schnirelmann apparently proved that any even number can be written as the sum of not more than 300,000 primes—a far cry from two primes. Ivan M. Vinogradov(1891-1983) proved that all sufficiently large odd integers are the sum of three primes. In 1973, Chen Jing-run showed that every sufficiently large even number is the sum of a prime and a number that is either prime or has two prime factors.

FERMAT'S LAST THEOREM

In the 1600's Pierre de Fermat (1601-1665) wrote in the margin of one of his books—

To divide a cube into two cubes, a fourth power, or in general any power whatever above the second, into two powers of the same denomination, is impossible, and I have assuredly found an wonderful proof of this, but the margin is too narrow to hold it.

restated: *If n is a natural number greater than 2, there are no positive whole numbers x, y, z such that $x^n + y^n = z^n$.* Fermat's note went out as a challenge. For centuries the proof or disproof has eluded even the most prominent mathematicians.

The next section gives additional background and discusses the latest information on Fermat's Last Theorem. Some of the discoveries that have been an outgrowth of efforts to prove Fermat's Last Theorem are probably more important than the theorem.

* * *

The study of mathematical ideas that have not been resolved is as interesting as delving into what we do know. This is but a small sample of unsolved mathematical mysteries. Although a few are simple enough to explain to someone with no mathematical background, their solutions are elusive.

---

[1]The *three impossible construction problems of antiquity*, which were to be solved using only a straightedge and compass, are: *trisecting an angle* (dividing an angle into three congruent angles), *duplicating a cube* (constructing a cube with twice the volume of a given cube), *squaring a circle* (constructing a square with the same area as a given circle). A few of the discoveries that these three problems helped evolve were the conchoid of Nicomedes, the spiral of Archimedes, and the quadratrix of Hippias.

[2]The challenge of the *Königsberg bridge problem* was to find a path over the seven bridges of Königsberg without crossing any bridge more than once. Euler developed the notion of networks when solving this problem.

[3]The *Parallel Postulate* dealt with determining whether Euclid's fifth postulate was indeed a postulate and not a theorem. Attempts to prove this postulate led to the discoveries of non-Euclidean geometries.

FERMAT'S LAST THEOREM

There are no positive whole numbers that solve $x^n + y^n = z^n$ when n is a natural number greater than 2.

When 17th century mathematician Pierre de Fermat scribbled this note in a margin of a translation of *Arithmetica* by Diophantus, little did he know the impact his comment would have on the development of mathematics for the next 350 years. Had he really solved it? Was he just playing a joke? No one will ever know for sure, but what we do know is that it became one of the famous unsolved problems in the history of mathematics. Like the *three famous construction problems of antiquity, the Königsberg bridge problem,* and *Euclid's 5th Postulate, Fermat's last theorem* has stimulated mathematical thought and discoveries for centuries.

ARITHMETICA

I have found a truly wonderful proof which this margin is too small to contain.

— Pierre de Fermat

The mathematical community is very excited and enthusiastic about *Modular Elliptic Curves and Fermat's Last Theorem,* the 200 page work of Andrew J. Wiles, professor of mathematics at Princeton University. Presenting his work at talks at Cambridge (June 1993), he climaxed his last talk with the announcement that he had proven the *Shimura-Taniyama-Weil conjecture,* which mathematicians felt held a key to proving Fermat's last theorem. In mathematical circles the feelings are very positive that Wiles' work has put *Fermat's Last Theorem* to rest.

Over the centuries there have been thousands of "proofs" of Fermat's problem, but none thus far has held up to scrutiny. It is not that Fermat's last theorem is so extraordinary, but as Fermat said "I have found a truly wonderful proof" — the solution (its proof) is the beauty of this theorem. It has sparked discoveries in number theory, cryptology and codes, to name some areas.

Andrew Wiles

Wiles had been intrigued with Fermat's theorem as a teenager. But did not delve into its proof until he saw a possible means. Wiles considers his proof a collaboration of the many mathematicians and efforts that went before him. Among historical plateaus toward this end we have 18th century mathematician Leonhard Euler who proved the theorem for n=3. German mathematician Ernst E. Kummer proved the theorem for all but three numbers less than 100.

Present day computer proofs have shown that no solutions exist for the first four million natural numbers. In the 1950's Yuktaka Taniyama made his conjecture involving elliptic curves and their structures in a hyperbolic plane. In the 1980's Gerhard Frey posed that if Taniyama's conjecture were true for a certain type of elliptic curves (called semistable), then Fermat's theorem could be proven. When Kenneth A. Ribet proved Frey's proposition, Wiles decided to undertake Fermat's theorem. From that time on he worked intensively for seven years. In May of 1993, a paper by Barry Mazur of Harvard came to Wiles' attention. The paper described a numerical technique that was over a hundred years old which was very important in the finalization of his proof.

GALILEO & PROPORTION

There are many mathematical concepts that have no restrictions in the realm of mathematics, but have limits when applied to the real world. The idea of proportion is one such idea that is a very useful tool in solving many types of problems. For example, if three identical boxes of marbles weigh 42 pounds, how many boxes are needed for 168 pounds? Setting up a proportion is one way to solve this— (3 boxes/42 lbs = ? boxes/168 lbs). But not all problems of proportion may be realistic to solve. Can any tree be scaled down to any size and be functional? Can a person of any size

exist? The composition of an object, be it a tree or the bones of a person, play vital roles in deciding the upper and lower limits of its size. A hundred foot person is not possible, since the structure and materials that make up the human body were not intended for such mammoth size. Even giant redwoods have limits to their heights, dictated by their roots system and the properties of wood. One of the first records of the problem of scaling an object up or down was tackled by Galileo in 1638 in his work *Dialogues Concerning Two New Sciences.*

Here he states "... *if one wants to maintain in a great giant the same proportion of limb as that found in an ordinary man he must either find harder and stronger material for making the bones, or he must admit a diminution of strength in comparison with men of medium stature; for if his height be increased inordinately he will fall and he crushed under his own weight. Whereas, if the size of a body be diminished, the strength of that body is not diminished in proportion; indeed the smaller the body the greater its relative strength.*"*

Dialogues Concerning Two New Sciences. Henry Crew & Alfonso De Salv translators. Macmillan, 1914.

MATHEMATICS & CONTAINERS

It is often startling and amazing to discover how objects and ideas were mathematized in the past. There are numerous examples of varied vases, containers, and vessels for storage that have been designed into many forms. This

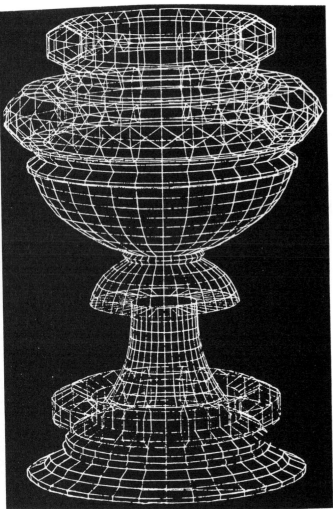

design and drawing of a chalice by Paolo Uccello is from the collection on display in the Uffizi Gallery of Florence, Italy. Although it was done during the first part of the 1400s, its precision and exactness is reminiscent of a computer analysis of a goblet, and illustrates linear perspective, constant ratios of proportional measurements, and use of geometric solids.

GEOMETRIES — OLD & NEW

I have discovered such wonderful things that I was amazed....out of nothing I have created a strange new universe. —János Bolyai from a letter to his father, 1823.

When most of us think of geometry, we think of our high school geometry course — all those "deadly" theorems we had to "memorize", the excitement of doing our first proof, points, lines, triangles, quadrilaterals, circles, solids, area, volume.   The beautiful thing about geometry is that one can visualize its elements. Little did we realize that a silent evolution of ideas had been taking place from ancient times to the present. Those of us who continued the study of mathematics learned that there was more, much more to geometry than Euclidean geometry. We learned how geometry and algebra were linked via the coordinate system by Rène Descartes. We saw how Euclid's 5th (Parallel) postulate was challenged by mathematicians for centuries. Many mathematicians felt the Parallel postulate was not an independent idea, but could be proven from the existing materials of geometry. As history has shown, the Parallel postulate definitely was independent in Euclidean geometry — but the unsuccessful attempts led to discoveries of non-Euclidean geometries. It is impossible to really separate one field of mathematics from the rest, for as mathematicians get ideas they rely on their entire mathematics knowledge. A time line of the evolution of geometries is too intriguing to pass up. Space limits this time line to an evolving of the fields of geometries rather than specific ideas within a particular geometry. Hopefully it will serve as a springboard for your own research.

600 BC—Thales introduces deductive geometry. It was developed over the years by such mathematicians and philosophers as Pythagoras, the Pythagoreans, Plato, Aristotle.

300 BC—Euclid compiles, organizes and systematizes geometric ideas, which had been discovered and proven, into thirteen books, called *The Elements.*

140 BC—Posedonius restates Euclid's 5th postulate.

3rd century AD—Proclus (410-495 AD) is one of the first recorded critics of Euclid's 5th postulate.

Countless attempts are made over the centuries to prove Euclid's 5th postulate.

1637—Renè Descartes formulates analytic geometry.

Gerolamo Saccheri (1667-1733) is the first to try an indirect proof of Euclid's Parallel postulate. Unfortunately, he does not accept the results of his work. Prior to his death he publishes a book, *Euclides ab omni naevo vindicatus (Euclid Freed of Every Flaw)*, which came to the attention of Eugenio Beltrami one and a half centuries later. Had Saccheri not rejected his findings he would have sped up the discovery of a non-Euclidean geometry by about a century.

1639—Girard Desargues (1591-1661) publishes a work on conics in which he discusses his new work on his discoveries of projective geometry.

1736—Leonhard Euler (1707-1783). His study and solution to the Königsberg bridge problem launches the field of topology.

1795—Gaspard Monge (1746-1818) describes structures by plane projections.

1822 Jean Victor Poncelet (1788-1867) revives projective geometry with his treatise and formulates the duality principle.

1843—Arthur Cayley (1821-1895) begins the study of n-dimensional spaces in analytic geometry.

Georg Cantor (1845-1918). His set theory provides a basis for topology, which is presented in 1895 by Henri Poincaré (1854-1912) in his *Analysis Situs.* Develops the Cantor set, an early fractal.

1871—Christian Felix Klein (1849-1925) does extensive work in projective geometry and topology, and proves consistency of Euclidean, elliptic & hyperbolic geometries.

19th century—Nicolai Lobachevsky (1793-1856), Jonas Bolyai (1802-1860), and Carl Gauss (1777-1855) independently discover hyperbolic geometry.

1854 —G.F.Bernhard Riemann (1826-1866) presents elliptical geometry.

1858 —August Möbius & Johann Listing independently discover one-sided surfaces, e.g. Möbius strip.

1888—Guiseppe Peano (1858-1932) creates Peano space-filling curve (fractal).

1904—Helge von Koch (1870-1924) creates Koch snowflake curve (fractal).

1919—Felix Hausdorff defines fractional dimensions in fractal geometry. A.S. Besicovitch generalizes Hausdorff's work.

1971—Vladimir Arnold links algebraic (n-dimensional analytic geometry) geometry and topology.

1951-75 —Benoit Mandelbrot, coins the word fractal and works nearly singlehandedly on its development.

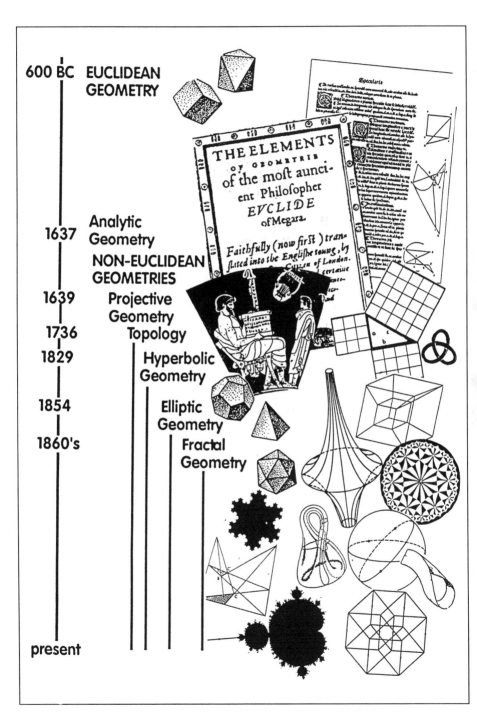

600 BC	EUCLIDEAN GEOMETRY
1637	Analytic Geometry
	NON-EUCLIDEAN GEOMETRIES
1639	Projective Geometry
1736	Topology
1829	Hyperbolic Geometry
1854	Elliptic Geometry
1860's	Fractal Geometry
present	

THE ELEMENTS OF GEOMETRIE of the most ancient Philosopher EVCLIDE of Megara.

Faithfully (now first) translated into the Englishe toung, by ... Citizen of London.

WHAT'S IN A NAME?
HYPERBOLIC GEOMETRY IS NOT NAMED AFTER THE HYPERBOLA

Ever wonder how certain mathematical fields get their names? For example, consider *hyperbolic* and *elliptic geometries*. In both cases their creators had nothing to do with the names that were eventually adopted. Hyperbolic geometry was discovered independently by both Nikolai Lobachevsky (1793-1856) and Janos Bolyai (1802-1860). Euclidean geometry's fifth postulate states that one and only one parallel line can pass through a given point P not on a given line L. Mathematicians unsuccessful efforts to show this postulate was provable, and thus a theorem, led to the discovery of non-Euclidean geometries. In *hyperbolic geometry* it was discovered that there exists more than one line passing through P and parallel to L. *Hyperbolic* comes from the Greek word *hyperbole*, meaning excessive. In this case it applies to the

Nikolai Lobachevsky was honored on a Russian stamp in 1958.

number of parallel lines passing through P and parallel to L. Lobachevsky had originally referred to his geometry as *imaginary geometry* and *pangeometry*. But the name by which it is called today — *hyperbolic geometry* —was given by the well known geometer Felix Klein, creator of the Klein bottle. In addition, Klein also coined the term *elliptic geometry* for Georg Riemann's non-Euclidean geometry in which no lines exist parallel to L passing through P. The Term *elliptic* is from the Greek term *elleiptis,* meaning to lack or fall short.

EULER'S MAGIC FORMULA —

The special thing about mathematical ideas is that once they have been established as true, they should hold for all cases. For example, to sum the first k counting numbers, 1+2+3+...+k, all one needs to do is plug into the formula k(k+1)/2. This formula has been proven mathematically by a method call induction.   It is physically impossible to test his formula for every possible set of consecutive counting numbers starting with 1, but the beauty of mathematical proofs is that they do not require brute force. Swiss mathematician Leonhard Euler is credited with many mathematical discoveries, especially in the field of topology. His solution to the Königsberg bridge problem is said to have launched the study of topological networks. Topology studies properties of objects that do not change as the object is distorted. For example, a cube can be distorted into a tetrahedron by stretching and squashing the cube, or vice versa. The size of the cube obviously changed, as did its number of faces, vertices or edges. As a result, one might wonder, what type of properties are left to observe

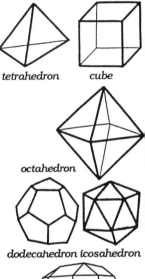

tetrahedron   cube

octahedron

dodecahedron icosahedron

rhombicosidodecahedron

and remain unchanged? One observation is that any point on the interior of the cube remains an interior point of the tetrahedron. Outside of topology, a fascinating theorem that Euler proved about a type of invariant property of polyhedra, is that *if you add the number of faces of the polyhedra, to the number of its vertices and then subtract the number of its edges, the result is* **always** *2.* **F+V-E=2.** Try it out on the Platonic solids shown. If you are feeling energetic try it on the rhombicosidodecahedron.

MATHEMATICS
PLAY ITS MUSIC

MATHEMATICS & MUSIC

MUSICAL SCALES & MATHEMATICS

MATHEMATICS & SOUND

> *Music is the pleasure the human soul*
> *experiences from counting without*
> *being aware it is counting.*
> **—Gottfried Wilhelm Leibniz**

Sounds, whether in the form of noises or music, are all caused by the vibrations of objects. Once objects, such as a a rubber band, a piece of wood, a wire, or a column of air in a flute begin to vibrate, they cause surrounding molecules of air to vibrate. These vibra-

tions travel outward from the source in three dimensions. When these vibrations reach our eardrums, the eardrums' vibrations send signals to our brains which create the sensation of hearing. Each musical instrument has a method of creating vibrations which in turn cause vibrations throughout the structure and material of the instrument. For example, when a guitar's string is plucked its vibrations cause the other strings and entire instrument to vibrate.

Since ancient times, mathematics has been used to explain music. Today computer modeling and digitization, the quantization of music and sounds coupled with the study of acoustics and acoustical architecture are producing new sounds sensations.

MATHEMATICS &
MUSIC

*May not Music be described
as the Mathematics of sense,
and Mathematics as the Music
of reason?* — **J.J.Sylvester**

Music and mathematics have been linked through the centuries.
During the medieval period the educational curriculum grouped
arithmetic, geometry, astronomy and music together. Today's
modern computers are perpetuating that tie.

Score writing is the first obvious area where mathematics reveals
its influence on music. In the musical script, we find tempo (4:4
time, 3:4 time, and so forth), beats per measure, whole notes, half

notes, quarter notes, eighth notes, sixteenth notes, and so on.
Writing music to fit x-number of notes per measure resembles the
process of finding a common denominator— the different length

notes must be made to fit a particular measure at a certain tempo. The composer creates music that fits so beautifully and effortlessly together in the rigid structure of the written score.   When a completed work is analyzed, every measure has the prescribed number of beats using the various desired lengths of notes.

In addition to the obvious connection of mathematics to the musical score, music is linked to ratios, exponential curves, periodic functions and computer science.

With ratios, the Pythagoreans (585-400 B.C.) were the first to associate music and mathematics. They discovered the connection between musical harmony and whole numbers by recognizing that the sound caused by a plucked string depended upon the length of the string. They also found that harmonious sounds were given off by equally taut strings whose lengths were in whole number ratios — in fact every harmonious combination of plucked strings could be expressed as a ratio of whole numbers. By increasing the length of the string of whole numbers ratios, an entire scale could be produced. For example, starting with a string that produces the note C, then 16/15 of C's length gives B, 6/5 of C's length gives A, 4/3 of C's gives G, 3/2 of C's gives F, 8/5 of C's gives E, 16/9 of C's gives D, and 2/1 of C's length gives low C.

Have you every wondered why a grand piano is shaped the way it is? Actually there are many instruments whose shapes and structures are linked to various mathematical concepts. Exponential functions and curves are such concepts. An exponential curve is one described by an equation of the form $y=k^x$, where $k>0$. An example is $y=2^x$. And its graph has the shape illustrated.

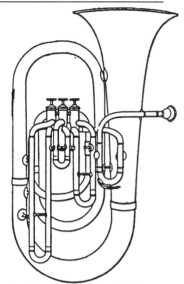

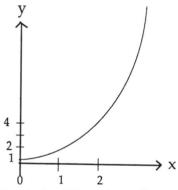

Musical instruments that are either string instruments or formed from columns of air, reflect the shape of an exponential curve in their structure.

The study of the nature of musical sounds reached its climax with the work of the 19th century mathematician John Fourier. He proved that all musical sounds — instrumental and vocal — could be described by mathematical expressions, which were the sums of simple periodic sine functions. Every sound has three qualities —

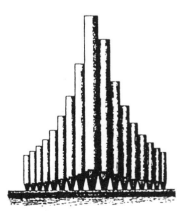

pitch, loudness and quality — which distinguishes it from other musical sounds.

Fourier's discovery makes it possible for these three properties of a sound to be graphically represented and distinct. Pitch is related to the frequency of the curve, loudness to the amplitude and quality to the shape of the periodic function[1].

Without an understanding of the mathematics of music, headway in using computers in musical composition and the design of instruments would not have been possible. Mathematical discoveries, namely periodic functions, were essential in the modern design of musical instruments and in the design of voice activated computers. Many instrument manufacturers compare the periodic sound graphs of their products to ideal graphs for these instruments. The fidelity of electronic musical reproduction is also closely tied to periodic graphs. Musicians and mathematicians will continue to play equally important roles in the production and reproduction of music.

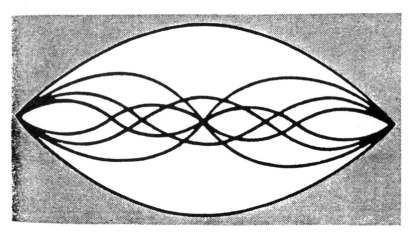

The diagram shows a string vibrating in sections and as a whole. The longest vibration determines the pitch and the smaller vibrations produce harmonics.

[1]A periodic function is one that repeats its shape at regular intervals.

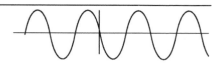

MUSIC SCALES & MATHEMATICS

The speed of light, c, π, e, φ and Avogadro's number are all examples of constants of our universe. These are numbers which have vital roles in equations and formulas which define various objects of our world — be they geometric, physical, chemical or commercial. Among these famous constants, the concept of an octave should be included as a constant of a special nature. The octave plays a vitally important part in the world of music. It establishes the unit or distance of a scale. Just as the ratio of a circle's circumference to its diameter always produces the constant π, the ratio of the

number of vibrations of a plucked string to a string half its length is the ratio 1/2. These notes have the same sound, and form the length of an octave.[1] The shortened string vibrates twice the amount per second than the original string. The number of notes or subdivisions of an octave is arbitrary, and is influenced by a number of factors. Consider the various factors that come into play when creating a particular scale. Sounds or notes are what

comprise a scale.   Each has a particular frequency.[2]   As
mentioned, two notes are an octave apart if the frequency of one is
double that of the other.[3] The trained ear can hear about 300
different sounds or notes in one octave.   But to produce a scale
with this many notes would be ludicrous, since traditional
instruments cannot produce that many notes.   For example, if
there were 300 notes in an octave, a piano of eight octaves would

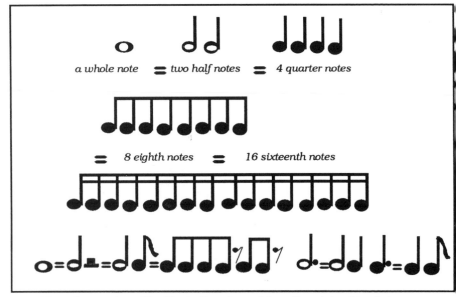

a whole note = two half notes = 4 quarter notes

= 8 eighth notes = 16 sixteenth notes

Musical equations. The illustration above shows the relaionships between
a whole note, a half note, a quarter note, an eighth note, and a sixteenth
note. A dotted note is one way of conveying fractions notes, because a dotted note equals one and one-half times its value.

have 2400 white keys. Can you imagine a pianist running back
and forth playing a keyboard this long? Thus, the number of
possible notes is restricted by the physiology of our ears and by the
capabilities of our instruments.   How and which of the 300
discernible sounds were selected for a scale? Selecting the notes
of a scale is analogous to selecting a numeration system. − What
base should be used and which symbols should be made to
represent the numbers? − With the scale, the length of an octave

string needed to be selected and the number of subdivisions (notes comprising the scale) had to be determined. As with numeration systems we find these evolved differently in various civilizations. The ancient Greeks used letters of their alphabet to represent the seven notes of their scale. These notes were grouped into tetrachords (four notes), which were put into groups called modes. The modes were the forerunners of modern Western major and minor scales. The Chinese used a pentatonic (five note) scale. In India, music was and is improvised within specific boundaries defined by ragas. This octave is divided into 66 intervals called srutis, although in practice there are only 22 srutis, from which two basic seven-note scales are formed.

The score of a 14th century missal page of a Gregorian chant.

The Persian scale divided the octave into either 17 or 22 notes. We see that although the octave was a given constant, different musical systems evolved. In addition, the musical instruments from one culture cannot necessarily be used to perform the music of another.

* * *

Archeological artifacts of instruments, vases, statues, and frescoes depicting vocal and instrumental musicians have been unearthed. There are many early examples of written music — Sumerian clay tablets excavated in Iraq appear to show an eight note scale (circa 1800 B.C.); fragments written on stone carvings and pieces of papyrus from ancient Greece, Greek text books (circa 100 A.D.); a Greek manuscript in which notes were written

A fresco depicting musicians from the tomb of Djeserkara in Thebes.

using their alphabet (circa 300 A.D.); manuscripts of a Moslem-Arabic chant from 8th century Spain.

During the 6th century B.C., Pythagoras and the Pythagoreans were the first to associate music and mathematics. The Pythagoreans believed that numbers, in some way, governed all things. Imagine their delight when they discovered the octave of a note, the periodicities of notes, and the ratio of notes on string instruments. In addition, they believed that planets had their own music, that the celestial bodies produced musical sounds. This idea came to be known as the "music of the spheres". Kepler believed in the music of the spheres, and in fact wrote music for each of the the known planets. Today, astronomers have received radio signals carried by solar winds. These sounds, including whistles, pops, whines, hisses when synthesized at increased speeds, become more melodious. Scientists have also observed oscillations from the sun which they surmise produce vibrations that are in various periods. Are musical scales needed to produce music? If they were, how

would birds sing? Yet most verbal renditions of a story or of a musical tune change slightly with each oral communication. To have a composition performed, scales are essential. Scales are the written language of music, even as equations and symbols are the written language of mathematics.

---

[1]The term octave comes from the Latin word for 8. A diatonic scale, has seven distinct notes from C to B and high C, the octave, is the eighth note.

[2]Frequency is the number of vibrations per second. There is a one-to-one correspondence between sounds and numbers (its frequencies), although our ears are not designed to discern all the possible distinct sounds.

[3]A plucked string produces a certain note, e.g. C has 264 vibrations per second. When the string is depressed at half its length, its octave is produced, which vibrates at 528 vibrations per second.

MATHEMATICS & SOUND

Mathematical ideas have been twisting and turning music and sound waves for centuries. A walk around the interior of the dome in St. Peter's Cathedral in Rome will convince you that the curve of the dome's walls carries one's whispers to a listener on the

St. Peter's Cathedral, Vatican.

opposite side.   Attending a Greek tragedy at the ancient amphitheater at Epidaurus illustrates that its designers must have studied and experimented with the mathematics of acoustics before designing and locating this phenomenal outdoor theater. A spectator seated in the farthest row can easily hear an actor drop a pin on the center of the stage. Specific mathematical shapes are used to design sound reflectors which are hung from the ceilings of symp̣ ̣ ̣ ̣ ̣ ̣ ̣ ̣ ̣ ̣ ̣ ̣ ̣ ̣ ̣ ̣ ̣ ̣ ̣Iall

For two parabolas situated as shown, sound originating at a focal point bounces off the parabolic ceiling, travels parallel to the opposite ceiling where it bounces to the other focus point.

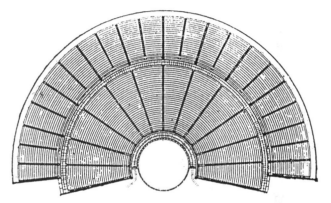

The above diagram is a schematic design of the ancient Greek amphitheater at Epidaurus, Greece. The photograph below is of the amphitheater today. Both the design and ideal location enhance its acoustics.

(south of the Rotunda in the United States Capitol building) reflects the conversations of individuals standing at focal points of the parabolas. Two people can carry on a normal conversation from the two focal points, undisturbed by the noise level of the hall.

As we see, finding ideal points for emitting and receiving sounds is not coincidental. Acoustics and sound are linked directly to mathematical objects and ideas. Sound waves themselves were shown to be simple periodic sine functions in the 19th century by mathematician John Fourier. He proved pitch, loudness and quality of sound were related respectively to the frequency, amplitude and shape of sinusoidal functions.

And most recently, QSound has been invented by Danny Lowe and John Lees, recording engineers and mathematicians. QSound produces sound in multi-dimensions. Unlike stereo which has different sounds coming from different speakers, QSound comes at you from all directions. You literally hear the recording in 3-dimensions. It requires no new equipment other than a stereo that plays a cassette or CD that has been imaged with QSound [1]. With such a recording just position yourself and your speakers as illustrated, and let the mathematics of sound do the rest.

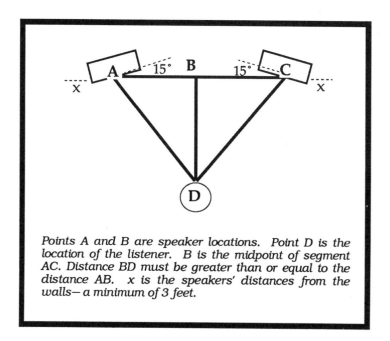

Points A and B are speaker locations. Point D is the
location of the listener. B is the midpoint of segment
AC. Distance BD must be greater than or equal to the
distance AB. x is the speakers' distances from the
walls— a minimum of 3 feet.

[1] As of February 1991, Madonna's *The Immaculate Collection* and Sting's
The Soul Cages were the only commercially available QSound albums.
Future releases by Janet Jackson, Paula Abdul, Bon Jovi, Europe, Richie
Zambora, Stevie Nicks, Julian Lennon, Freddie Jackson, and Wilson
Phillips are scheduled to use the patented process.

0111000**BAUD**11010

1**BINARY** 1**CODE**001

BINARY 0**NUMBERS**

0110101**BIT**100**BYT**

1010111**BUG**011001

01**CHIP**010**FORMAT**

100**HARDWARE**010

0**SOFTWARE**110001

FLOPPY0**DISK**01011

1001**INPUT**00**RAM**

01**ROM**1010**MOUSE**

BUS010**KILOBYTE**11

0**ASCII**0110**SCSI**0110

0**WYSIWYG**10**VIRUS**

111**PIXEL**01101**BOOT**

MEGABYTE011**PORT**

THE COMPUTER REVOLUTION

THE COMPUTER, 21ST CENTURY'S TOOL FOR CREATIVITY

> *To err is human, but to really foul things up*
> *requires a computer. —***Anonymous**

Whether we are ready or not or whether we like it or not, the computer is the tool of the 21st century.   The computer is impacting all facets of our lives both positively and negatively.   It has sped up the gears of change. For today's writers the computer can be the pen, pencil, or typewriter. For the accountant or bank teller it can be the calculator and keyboard. For the artist — a paintbrush and palette, for the scientist a laboratory, for the architect the drafting table, for the engineer design tools, for the professor a research tool, for the librarian a card catalog and reference desk, for the mathematician an incredibly fast and accurate calculator/computer. However, when things don't go exactly as planned, the computer becomes the scapegoat, and we are told— "it lost your file," "the system is down," "there seems to be a glitch." Nevertheless, since they have touched almost every aspect of our daily lives, we are now dependent on computers.

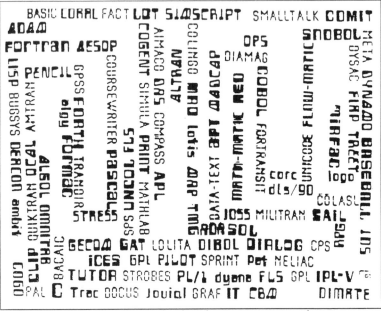

Some computers languages developed to communicate with computers.

A LOOK AT THE PAST

OBSOLETE CALCULATORS

•The ten digits of our hands were our earliest counting device.

•The Chinese devised a segmented box to use with their rod numerals. This box was used to write systems of equations.

•The abacus was used in calculation by many cultures including the Chinese, the Greeks, the Romans, and the Japanese.

•The Incas used the knotted quipu as their accounting device.

•Napier's rods were invented by John Napier in the 1600s to aid computation.

•The slide rule was invented around 1620 by Edmound Gunter.

•The first adding machine was invented by Blaise Pascal in 1642. And in 1673, Gothfired Wilheim von Leibniz invented one that could also multiply and divide.

•In the early 1800s Charles Babbage's designs and work on the difference and analytical engine furnished the foundation for the modern computer.

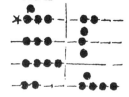

ADDITION.

Master.

The easiest way in this arte, is to adde but two summes at ones together: how be it, you maye adde more, as I wil tel you anone. therefore whenne you wylle adde two summes, you shall firste set downe one of them, it forceth not whiche, and then by it drawe a lyne crosse the other lynes. And afterwarde sette doune the other summe, so that that lyne maye be betwene them; as if you woulde adde 2659 to 8342, you must set your summes as you see here.

And then if you lyst, you maye adde the one to the other in the same place, or els you map adde them bothe togither in a newe place: which way, bycause it is most plyneste

This illustration is a reproduction from the early English arithmetic book **The Grounde of Artes**, by Robert Recorde. In addition to the lines representing 0s, 10s, 100s, etc., the places between these lines were also used to represent 5s, 50s, 500s following the Roman numerals. The "x" on the line was initially used to mark the 1000s line, but later it was used to indicate a comma when writing such numbers as 23,650. When five counters accumulated on a line those beads would be removed and one counter would be **carried** to the above space. Hence, the possible origin of the term "carry". In addition to writing this book, Robert Recorde (circa 1510-1558) introduced the symbol "=" for equality, wrote an important algebra text book called **The Whetstone of Witte** and the geometry book, **Pathway to Knowledge**.

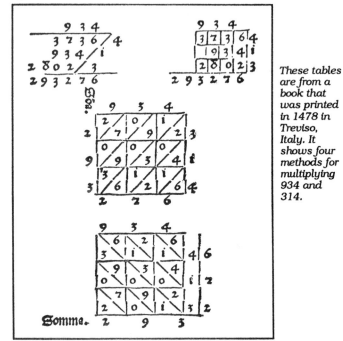

These tables are from a book that was printed in 1478 in Treviso, Italy. It shows four methods for multiplying 934 and 314.

The binary system (base two), which uses only 0s and 1s to write its numbers, held the key to communicating with electronic computers — since 0s and 1s could indicate the "off" and "on" position of electricity. The famous Scottish mathematician John Napier (1550-1617) utilized the concepts of base two before the advent of electricity. Napier is best known for revolutionizing computation by his invention of logarithms. His calculating rods, known as Napier's rods (or bones) were based on logarithms and were used by merchants to perform multiplication, division, and could be used to find square roots and cube roots. Less well known is his chess board method of calculating. Although he did not use binary notation to write numbers, the board does illustrate how he expressed the numbers in base two. For example, to add 74 + 99+ 46, each number is written out in a row of the chess board by

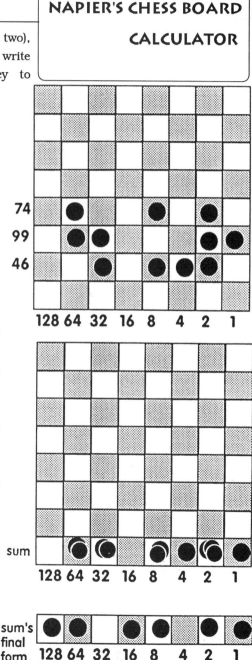

NAPIER'S CHESS BOARD CALCULATOR

74
99
46

128 64 32 16 8 4 2 1

sum

128 64 32 16 8 4 2 1

sum's final form

128 64 32 16 8 4 2 1

placing markers in the appropriate squares of the row so that the sum of the markers values (indicated along the bottom line) total the number they represent. 74 has markers at 64, 8 and 2 since 64+8+2=74. After each number is expressed on the chess board, the numbers are added by gathering the markers' vertically down on the bottom row. Two markers sharing the same square equal one marker to their immediate left. So two "2" markers, produce one "4" marker. Working from right to left, any two markers sharing the same square are removed and replaced by one marker in the adjacent square at the left. At the end of this process, no square will have more than one marker. The sum of the values of the remaining markers represents the sum of the numbers.

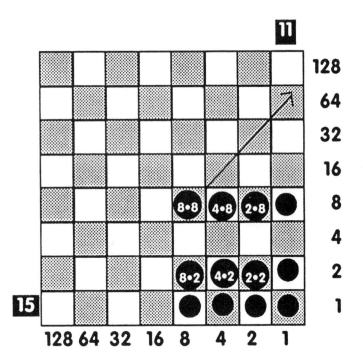

To multiply using Napier's chess board calculator, the number representations along the horizontal and vertical column are used.

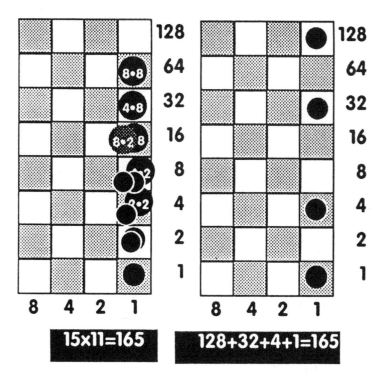

15x11=165   128+32+4+1=165

Suppose we want to multiply 15 x11. One number is expressed with markers along the bottom row and the other along the vertical column at the right. Then a new marker is placed at the intersecting square where a row with a marker meets a column with a marker. After this is done, the multiplication process is simply done by sliding the markers from the bottom row diagonally to the vertical row. As with addition, any place two markers occupy the same square, they are removed and one marker is placed in the square above them. The resulting vertical column will represent the product of 15 x11.

A LOOK AT THE PRESENT

COMPUTERS ARE IN OUR TREES

It is unworthy of excellent men to lose hours like slaves in the labor of calculation which could safely be relegated to anyone else if machines were used. —**Gottfried Wilhelm von Leibnitz**

"I'm a dinosaur." "I refuse to have anything to do with them." "I can't understand how a person can spend so much time in front of one." "They're so impersonal." "I'm just computer illiterate." How many times have you heard or perhaps made some of these statements ? But it must be accepted, regardless of how some people may feel, that computers are here to stay. They are making our lives easier in some ways, more complicated in others, and invading our privacy. In fact, it is difficult to remember the days before computers. Now computers are even in our trees!

The growth of trees can be patterned using mathematical networks or drawn using fractals. Computers can be used to model forest fires, and thereby develop strategies to either put out fires or create controlled burns. In addition, computers are now being used extensively to log urban forests. Many cities, in efforts to preserve and maintain the health of their trees are resorting to computers. Washington D.C. has a database of about 109,000 street trees, while Paris, France also has its 100,000 trees on computer. What kind of information is entered? Each city usually decides what data is essential for its particular situation. The Paris database includes— the tree's location (each tree is consecutively numbered along the street, and its distance from the building, curb and adjacent trees are noted), its vital statistics (species, sex,

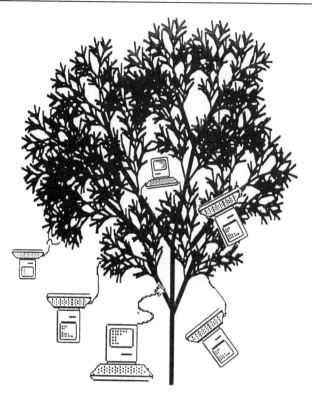

age, trunk size, height), type of pruning received, environmental statistics (includes soil type, planter dimensions, drainage, mulching), tree's health (status and diseases noted), environmental pollutants and effects. The initial logging of such information is very time consuming, as is the updating of files. Since trees are an invaluable asset to any community, both aesthetically and environmentally, the time invested and long range outcome make such databases worthwhile. In San Francisco, California, the city gardeners maintain the city's inventory using laptop computers.   Paris followed a method used in Washington D.C., in which the initial inventorying was done by junior forestry engineers using hand-held computers. Without computers, such sophisticated databases would not have been possible. Storage and filing space would have been too formidable, let alone the necessary updating, searching and sorting of data.

* * *

As the computer evolved, computer functions slowly invaded aspects of our lives. Even as early merchants and navigators, wondered how they had functioned without a set of Napier rods, and Inca scribes must have speculated how they had kept inventory of the population and products of the Inca empire without the quipu, so today we are at a similar point.

Bar codes for labeling & tracking such things as trees, groceries and packages.

As we see with the urban forest, computers play an incredible role in storing, processing, and retrieving data. Today, data imput devices are used by restaurants to process orders, bar codes are used by stores and businesses for sales, inventory, and tracking. Computers are invaluable in the sciences for analyzing, comparing and computing information, thereby saving tremendous amounts of time with fewer errors. This evolving tool's influence on civilization will be as great as was the impact of the electric light bulb.

MATHEMATICS BECOMES A PRIVATE EYE

Mathematical wavelets are a new crime fighting technique and tool being used in the United States. Mathematicians have developed wavelets analysis to process images in a quicker more accurate fashion. The method divides an image into basic components. It then isolates the important and relevant information needed to

reconstruct the image and discards the non-essential information. Consequently, wavelets reduce the amount of data necessary to store, transmit and retrieve the image. This data compression saves time and money. In addition, the wavelets do not distort the image as other methods do. This allows the FBI to more efficiently process, identify and match the image of a person's fingerprints.

WHAT'S MY SECRET ?

Governments, revolutionaries and financiers have always needed to send and receive top secrets. Many methods of coding and decoding messages have been devised over the centuries. Among these are the Spartan scytale[1], Polybius' square cipher[2], Julius Caesar's letter substitution, Thomas Jefferson's cipher wheel, and the

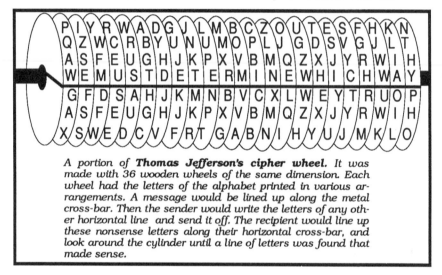

A portion of **Thomas Jefferson's cipher wheel.** It was made with 36 wooden wheels of the same dimension. Each wheel had the letters of the alphabet printed in various arrangements. A message would be lined up along the metal cross-bar. Then the sender would write the letters of any other horizontal line and send it off. The recipient would line up these nonsense letters along their horizontal cross-bar, and look around the cylinder until a line of letters was found that made sense.

German military code *Enigma*. There are also many examples in literature where cryptography is important to the plot, as in Edgar Allan Poe's *The Gold Bug*, Jules Verne's *La Jangada*, Conan Doyle's *The Adventure of the Dancing Men*.

For the past twenty years the encryption formula, DES, has been used by banks, government agencies and private companies to secure information. This method uses 56 bits of computer data which theoretically would take 200 years to decode using a super computer. Intelligence and law enforcement agencies fear that the present methods are not adequate and would not be able to intercept communications from criminals, terrorist and foreign governments. Consequently, a new system is being proposed with a new formula using 80 bits that would

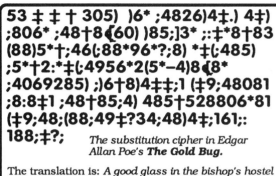

53 ‡ ‡ † 305))6* ;4826)4‡.) 4‡)
;806* ;48†8(60))85;]3* ;:‡*8†83
(88)5*†;46(;88*96*?;8) *‡(;485)
;5*†2:*‡(;4956*2(5*−4)8(8*
;4069285) ;)6†8)4‡‡;1 (‡9;48081
;8:8‡1 ;48†85;4) 485†528806*81
(‡9;48;(88;49‡?34;48)4‡;161;:
188;‡?;

*The substitution cipher in Edgar Allan Poe's **The Gold Bug**.*

The translation is: *A good glass in the bishop's hostel in the devil's seat twenty-one degrees and thirteen minutes northeast and by north main branch seventh limb east side shoot from the left eye of the death's-head a bee line from the tree through the shot fifty feet out.*

theoretically take more than one billion years to crack. This new approach uses a special microchip which would be inserted in telephones, satellites, fax machines, modems etc. to scramble secret communications. The chip would use classified mathematical formulas developed by the National Security Agency. In addition, mathematical keys that unscramble the coded data would be kept by the FBI and other agencies determined by the Attorney General of the United States. The manufacturing process of the chip makes it *nearly* impossible to take apart and decode. There are people, such as Mitchell Kapor of the Electronic Frontier Foundation (a public policy group in Washington D.C.), who say "A system based on classified secret technology will not and should not gain the confidence of the American public"[3]

---

[1]Plutarch describes how messages were exchanged by winding a narrow band of parchment or leather in a spiral manner around a cylindrical shaft. A message would then be written, and the band removed and sent to the recipient who had a cylindrical shaft of the exact same size to wind the band around and decipher the message.

[2](Circa200 B.C.-118 B.C.) the Greek Polybius devised a substitution cipher designed around a 5x5 square, which converted each letter of the alphabet into a two digit number using only the digits for 1 through 5.

[3]*Secret Phone Scheme Under Fire*, by Don Clark, San Francisco Chronicle, May, 1993.

PICKING OUT PRIMES

2 — 8 — 756839

4
5
9
11
14
23
37
6
18
51
87
641
273

One of the earliest methods for finding prime numbers was devised by the Greek mathematician Eratosthenes (275-194 B.C.). He created a numerical sieve that would eliminate multiples of numbers up to some given number. Ever since then mathematicians have been devising new means for seeking out and identifying prime numbers.

In 1640 Pierre de Fermat claimed that all the numbers of the form $F_n = 2^{2^n} + 1$ (for n=0, 1, 2,3,...) were prime. This is true for the first five Fermat numbers (for n=0,1,2,3 and 4). But the following century Leonhard Euler factored the 6th Fermat number (F_5) into 641 times 6700417. Then in 1880 the Fermat number, F_6, was factored into 274177 times 67280421310721. Currently no other Fermat numbers have been found to be prime, and as of 1993 the 22nd Femat number (F_{22}) was shown to be composite.[1] In 1644, the French monk Marin Mersenne wrote an expression— 2^P-1, with p a prime number— that would yield primes. But not all the numbers produced by this expression are prime. There are currently 33 known Mersenne primes. Working with expressions (such as those for Mersenne numbers, Fermat numbers, Carmichael numbers, and others), number theory, and computer programming techniques, mathematicians using either supercomputers or strings of revved-up personal computers seek out larger and larger prime numbers — and look for quirky characteristics or patterns.

WHY ALL THE INTEREST IN PRIMES? — *mathematical curiosity — setting new records — testing the efficiency and hardware of new computers— using primes numbers to form multidigit numbers for encoding confidential material.* Today's computers, along with ingenious programming and problem solving have made it possible to find literally enormous prime numbers. WHAT'S THE CURRENTLY LARGEST EXPLICIT PRIME? 2^{859433}-1.[2]

---

[1] *Projects in Scientific Computation* by Richard E. Crandall. Springer-Verlag, Santa Clara, CA 1994.

[2] Found by David Slowinski in 1993. Verified at his request by Richard Crandall.

7455560282564788420883739573620045491878336663342657

CRYPTOGRAPHY,
ANARCHY, CYBERPUNKS,
& REMAILERS

Ever wonder, irritatedly, how it is you receive so much junk mail — catalogues you never requested, offers for "super" buys? How were your name and address accessed? Welcome to the electronic age and the loss of privacy. Most individuals have considerable electronic data about them, which can be accessed by a computer/modem user to create a profile of you based on things you have bought, your travel tickets, your medical records, your traffic tickets, outstanding loans, etc. Although the loss of privacy was created by today's new technological methods for storing and accessing information, perhaps there is a way to restore that privacy by using these same methods. Cyberpunks, as they call themselves, have come to the rescue. They advocate restoring the privacy of the individual, by using sophisticated cryptographic methods to encode one's information, and hinder its ready availability. Naturally there are pros and cons to this cryptographic anarchy. Some feel that government and law enforcement agencies have a right to eavesdrop, and cyberpunks new methods would present obstacles. *Remailer* is an example of one new method developed using advanced cryptography which allows one to send information via modem without leaving any "footprints" to trace the sender. Many feel that if governments can use methods and devices to encode important information, an individual also has a right to use similar methods to insure personal privacy.

COMPUTERS, IRRIGATION & WATER CONSERVATION

It is eerie to observe lush crops growing from dry, sometimes cracked earth. But now with the use of computerized subsurface drip irrigation, water, fertilizer and occasionally pesticides can be delivered directly to the roots of plants. With subsurface drip ir-

rigation, raisin grape farmer Lee Simpson of Fresno, California has cut his water consumption in half and doubled the yield of his farm while cutting back the use of pesticides to one-sixth. A California Department of Water Resource test project on a 160-acre cotton field of Harris Farms, also in Fresno, produced 1600 pounds of cotton per acre as opposed to 930 pounds per acre from previous years, and used 7 inches less water per acre. Claude

Phené has been at the cutting edge of this "new" type of farming (drip irrigation has actually been around for about 20 years). Phené had been advocating conserving water and improving yields by this method for years, but it was not until 1987 that some farmers took notice. Phené tested his methods on his tomato crops in California. Acres that formerly produced 26 tons per acre, now produced 100 tons per acre. Phené points out that it does not make sense to flood a room to water a potted plant. His statistics claim that California agriculture uses 85% of the state's water supply, and if subsurface computerized drip irrigation[1] were used with just one crop— e.g. cotton— the 6 inches of water saved on each of the 1.4 million acres would be enough water to supply Los Angeles. The method also reduces the use of herbicides, since weeds do not grow as readily between the furrows because plants are watered at the roots and not flooded. In addition, the amounts of fertilizers and pesticides used would be cut by more than half. His computerized techniques allowed his operation to discover the interaction between different nutrients, timing of nutrient applications and their affects on crops yields, the optimum amount of fertilizer for optimizing plant quality, and numerous factors from soil texture to rooting characteristics.

---

[1]Crops are irrigated by underground pipes placed 18 inches below the surface.   The very sensitive lysimeters detect morning dew, measure amounts of water that plants absorb and emit, measure surface weather conditions.   The computer gathers and analyzes data, and delivers minute amounts of water throughout the day and occasionally fertilizer, herbicides, and pesticides. Without investing in a computer system, small farms could monitor their fields via modems at a centralized computer location.

COMPUTERS FIGHT FOREST FIRES

Today's computer modeling is a very powerful tool, used by scientists and professionals across a spectrum of fields. It can be used by economists to predicts economic cycles, by doctors to monitor and predict the spread of a contagious disease or with chaos theory to describe heart arrythmias. It has been used by sociologists along with statistics to observe a social trend. The list seems almost unending.

Not too long ago firefighters' tools were protective garments, axes, pike poles, ropes, chain saws, fireproof blankets, water, chemicals. Today some firefighters are also equipped with laptop computers or even a tent/laboratory in the middle of the forest complete with personal computers. In addition to keeping track of people and supplies, the computers are also used for forest fire analysis.

While working for the United States Forest Service's Inter-mountain Fire Sciences Laboratory in Montana, mathematician Patricia Andrews developed the program *Behave* in 1984. The location of the fire, the typography of the area, the climate conditions (wind speed and direction, dryness, etc.), the types of burning trees or grasses and much more information are fed into the computer. The program then predicts how the fire "might" behave, thereby helping fire managers decide how to best fight the blaze. Naturally, it cannot predict all possible outcomes, but *Behave* can be continually modified to become more comprehensive as new predicaments occur, such as the Yellowstone "crown fires" in which the fires spread from treetop to treetop. *Effects* is a companion program now being used to help forest management with decisions for controlled burns. These high tech firefighting tools are now being used in China, have been translated into Spanish, and information has been requested by Italy.

A LOOK AT THE FUTURE

> CYBERSPACE
> VIRTUAL REALITY

*It is the perennial youthfulness
of mathematics itself
which marks it off with a
disconcerting immortality
from the other sciences.*
—Eric Temple Bell

In the 1500s people were treated to the marvels of the camera obscura. Actual moving scenes from things taking place outside a room were projected on the wall of a darkened room. No electronic devices were needed to bring these scenes into the room. Then, in the second half of the 19th century, there was a surge of interest in optical illusions. Physicists and psychologists studied and wrote about how our minds were tricked by what we perceived. The physical structure of the eye and the analysis of how our minds process information from the eye were studied in efforts to explain distortions which tricked our minds into believing they existed.

Some of their findings were—

(1) The placement of certain angles and segments can lead our eyes inward and outward and thereby make an object appear to be shorter or longer.

(2) Horizontal objects have a tendency to appear shorter because the retinas of our eyes are curved.

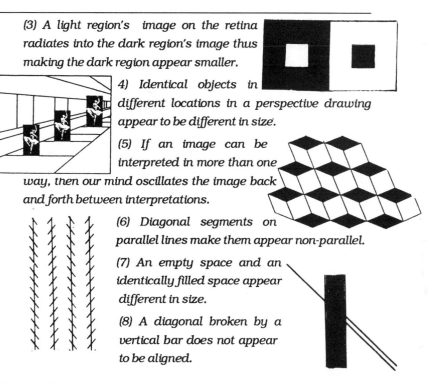

(3) A light region's image on the retina radiates into the dark region's image thus making the dark region appear smaller.

4) Identical objects in different locations in a perspective drawing appear to be different in size.

(5) If an image can be interpreted in more than one way, then our mind oscillates the image back and forth between interpretations.

(6) Diagonal segments on parallel lines make them appear non-parallel.

(7) An empty space and an identically filled space appear different in size.

(8) A diagonal broken by a vertical bar does not appear to be aligned.

Now, in the 20th century, computer scientists, mathematicians and inventors are carrying optics, computer technology and optical illusions to new heights with the **creation of artificial worlds.** Equipped with various paraphernalia, the observer is no longer simply an observer— but **actually enters worlds created by the computer.**

In these artificial worlds, one can be a participant. For example, one may choose to be:

— an "Olympic " runner experiencing the thrill of running for the gold metal

— an air traffic controller directing planes flying around in the 3-dimensions

— a weather forecaster flying around the world experiencing first hand weather conditions that have been programmed and processed into the computer.

— an atom about to be bonded with other atoms into a molecule

— an architect viewing his or her most recent design by actually walking through the computerized image room by room.

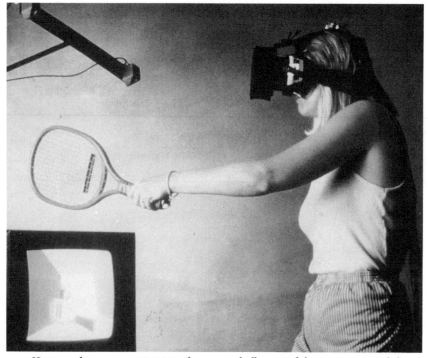

Here a cyberspacenaut enters the racquetball court of the computer, and the game begins. The players actually feel they are playing on the computer court. Photograph courtesy of Autodesk, Inc. Sausalito, California.

Virtual reality, cyberspace, and *artificial reality* are a few of the expressions coined to describe this new form of optical illusions. In the confines of a small room one can don computer gear,[1] and suddenly find oneself walking in the English countryside, inspecting a project's progress thousands of miles away, or learning how a bee collects pollen by becoming the bee. The

Computer generated living room awaiting a visitor to virtual reality.
Photograph courtesy of Autodesk, Inc. Sausalito, California.

applications of such technology are mind boggling. Cyberspace is
still in its initial stages with much yet to be developed and
refined[2]. Hopefully its evolution and popularity will not serve as a
means of mind control, but rather of mind expansion. Computer
scientists and mathematicians are breaking new ground in
computer graphics, especially by using fractal geometry to create
these special effects. Perhaps the holodeck of *Star Trek— The Next
Generation* is not so far fetched.

---

[1] The computer gear comes in many forms which can include special eye
gear, data gloves, data body suit.

[2] A few of the universities and businesses involved in these artificial
worlds are University of North Carolina; Autodesk, Inc. of Sausalito, CA;
HITL at the University of Washington; VPL Research Co. of Redwood City,
CA; Carnegie Mellon University. Connecticut State Museum of Natural
History has an installation created by Myron Krueger called Video Place.
At U.C. Berkeley's ASUC, one can play *Dactyl Nightmare*, a virtual reality
game.

HYPERTEXT

Speculations on the fourth dimension surfaced in the 19th century
when August Möbius noticed that
a shadow of a right hand could be
made into a left handed shadow simply by passing one's hand
through the third dimension. Little did one realize that the term
hypercube[1] would lead to a generation of such terms as
hyperspace, hyperbeing, hypercard[2] and now hypertext. Though
the last two terms are not directly related to the 4th-dimension,
they *are* related to the computer and its ability to jump from one
idea to another − interactive computing. It can be thought of as
analogous to moving from one dimension to another.   The
participants decide where to go and the computer transports them
there. For example, with interactive computing, after you decide
what you want to do, read or view, the computer immediately
brings before you information on that topic, which may include a
sound tract, graphics and even a video. Suppose you want to
learn something about V-E day of World War II. The computer is
able to present historical information, maps showing the invasion
and how it proceeded, newsreels from that period, and even songs
that were popular during that time. You choose topics by clicking
your choice[3].

Hypertext draws on interactive computing by allowing even more
active participation. A hypertext is not read from beginning to end
in the traditional manner. Instead, the computer functions as a
tool to explore other possible outcomes, and thereby a new story
evolves. With hypertext the reader spins his/her own tale by clicking on highlighted ideas that are interconnected in a story called
a web, which has been spun into the original program. While
reading the story on the computer, you can choose the direction in
which you want the story line to move. You do not create the
storyline, but rather choose options along the way, and see what

they unfold. A click on a key word or image may take you to a new location, a new idea or even a new plot line. You are able to see how the author unfolded the story along the paths you chose. The

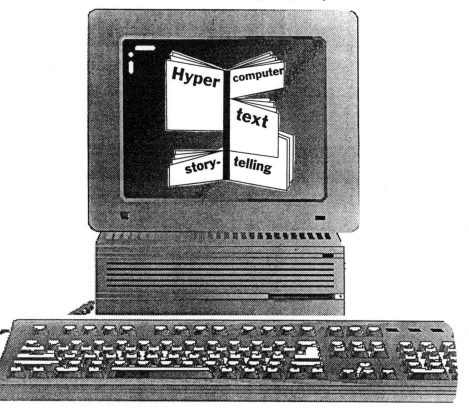

reactions to interactive literature are mixed. Some feel it is a novel curiosity that is far more interesting in theory than practice. It is in its initial stages, and thus far once a story has been "altered" by the reader, it cannot be undone. In addition, it may also be difficult for the novice to tell when the story is finished, since there may be key words or phrases available throughout the web allowing the reader to continually go off to something new or perhaps even around in a circle by mistake. The web stories may require time and/or practice for inexperienced hypertext readers to orient themselves to this new form of reading.   Improved pro-gramming and writing techniques for this new media need to be re-fined if it is to work optimally and if the reader is to have unencumbered freedom to explore its potentials.   As English professor George P. Landow of Brown University says,"It really has the potential to be the next wave of storytelling...The next question is, is this total chaos and anarchy, or is it a new reading form that makes the reader a kind of creator?"[4] There are many who feel hypertext is introducing a new way of writing, and consider it an art form. Most of the focus thus far has been on the reader who now helps shape the direction of the story. But consider what such story writing entails for the author. The writer does not develop a single plot line, but instead a family of possible plot lines and their relative outcomes, called the web. As with interactive computing, the web can be enhanced by computer graphics, sound and video. It is too early to pass judgement on interactive literature, especially since the hypertext stories are just being re-leased or posted on electronic bulletin boards, but it will be fascinating to follow the outcome . Will it catch on as a new craze? One thing is certain, hypertext would not exist if it were not for the modern computer.

[1] Hypercube is the term coined for a 4th-dimensional cube.

[2] Hypercard has two meanings. It can be the trade name for a computer program, which allows you to access and create stacks of cards with information in various forms (including graphics, video, sound) simply by clicking on a particular topic. Hypercard can also be a card (which is 2-dimensional) that has been cut and folded into a 3-dimensional model, similar to the book illustrated on the computer screen. It is hard for one to pass up trying to recreate a hypercard from a 3x5 file card.

[3] It seems an easy way to pursue learning. What are some drawbacks? You have to depend on the material that is accumulated in that particular piece of software. Sources other than those on that program are not at hand. If you are using the program for more than just general knowledge, you have to be careful not to rely on this as your sole research material. Hopefully this form of available information will not hinder creativity or the ambition to do additional research, but rather stimulate it.

[4] *A New Way To Tell Stories*, San Francisco Chronicle, San Francisco, CA, April 1993.

LITTLE FERMAT

One often gets the feeling that the layperson now feels that computers have evolved to their ultimate state, but computer scientists, mathematicians and scientists know this isn't so. In the spirit of Charles Babbage, M.M.(Monty) Denneau, George V. and David V. Chudnovsky and Saed G. Younis created Little Fermat, a

The keyboard punch for the 1890 Herman Hollerith tabulating machine, which revolutionized the procedure for taking the census in the United States.

computer designed to handle gigantic computation problems without the number errors associated with conventional computers. Using ideas from number theory—specifically modular arithmetic and Fermat numbers—the computer can carry out computations virtually error free. Little Fermat is programmed in a language called Younis. By utilizing Fermat numbers as divisors in modular arithmetic, it can speed up certain kinds of calculations and avoid the use of real numbers. Thus far Little Fermat is a one-of-a-kind computer, but its creators feel it is ideally suited for digital signal and imaging processes and the solving of various problems from hydrodynamics, chemistry and aerodynamics that require differential equations. Furthermore, they hope it will serve as a model to enhance the performance of the supercomputers.

COMPUTERS & A-LIFE

With technology changing at such a rapid rate and with new ideas and applications springing-up overnight, computers seem to be impacting every aspect of our lives whether we are aware of it or not. Many scientific laboratories are now the computer, utilizing its modeling and simulation abilities.

A-life, as it is often referred to, is a way to simulate lifelike forms, their behavior, reproduction and evolution. One recent example of its uses were the lifelike computer generated bats in *Batman Returns*. How was it done? Utilizing logic, mathematics and computers, the habits of a living thing, such as a bat, are analyzed into basic logical steps. These steps are developed into a computer simulation which captures the movements and habits of the life form. The applications are far reaching. One example is the research in disease done at Scripp Research Institute. Here Gerald Joyce uses computer simulations and develops a laboratory procedure in which enzymes evolve according to how they slice in the genetic code of an AIDS virus. By trial and error, these enzymes explore a cure. Computers can develop their own programs based on how well they solve certain problems. Programs designed by this method include steering a vehicle, predicting economic outcomes, predicting planetary motion. Computer pioneer, John von Neuman, contended that machines could not only process information, but reproduce themselves. Some proponents of A-life feel that the essence of life is a set of rules directing the interaction of cells, atoms, molecules, etc. A close connection between A-life and fractal mathematics is also developing. For example, computer simulations have been developed of plant cells that grow and divide according to a

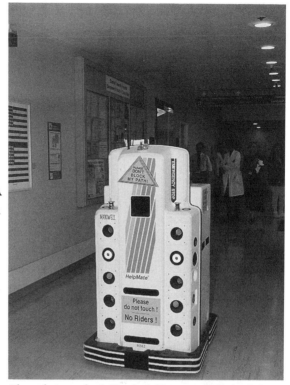

The robot used at Stanford University Hospital (Stanford, California) to carry records back and forth. Photograph courtesy of Stanford University Medical Center, Visual Arts Services.

set of instructions. One such grew into cell formations almost identical to those of a fern. Studying how different insects such as ants navigate, simulations of miniature robots have been created. These robots are instructed to move and change direction only when they come to an obstacle they cannot traverse. Such robots outperform the traditional larger designs when it comes to finding their way around obstacles. In fact, *Atitila*, designed by MIT's Mobot Lab, weighs 3.6 pounds and will hopefully be used to explore Mars.

OPTICAL COMPUTERS

Our present computers run on electricity. But some scientists are working on the development of a new type of computer that is made up of optical fibers, lasers and switches that process data and perform calculations via light[1] rather than electricity. Unlike electronic computers in which data and programs are stored either on

memory chips, hard drives, or disks the optical computer circulates data as light pulses through the optical fibers. "For the first time, we have a computer in which the program and data are always on the move in the form of light, eliminating the need for static storage," says Harry F. Jordan of the Optoelectronic Computing Systems Center at the University of Colorado at Boulder, Colorado. Over three miles of optic fiber serve as the computers main memory, where instructions and data are encoded in light pulses and circulate the optical fibers. Jordan and Vincent P. Heuring head the development at the University of Colorado. The computer that they have developed thus far does "demonstrates the principle that all of the components of a general-purpose machine can be done in optics."[2]

---

[1] The only time light is not used occurs when the optical switches are activated and light pulses momentarily converted to electrical current.

[2] Harry Jordan, quoted from *Science News* , January 23, 1993.

FUZZY LOGIC & COMPUTERS

Suppose I take a log and add it to the fire in my wood burning stove. It begins to burn immediately. At what precise moment will it no longer be considered a log? One person might say right after it caught fire. Someone else might feel when half of it is burned away, and yet another person might contend it's a log until its last ember. No matter how logical one tries to be about answering this question, there is no definite answer. There is no right or wrong answer. There is no way to quantify the answer. It is all a matter of degree.   A 5:00 pm appointment means the same for everyone.   But an appointment set for late afternoon may mean 3:30 pm to one person, 5:30 pm to another. In fact, there are a host of possible times depending upon whom you asked. And such is the case with so many things in life, for there is a subjective nature to life and the universe. True or false, yes or no logic cannot deal with such situations or with the ever changing status of everything on earth and in the universe. These are the things that *fuzzy logic* deals with.

Fuzzy logic is a new way of looking at things and analyzing the world. The

Fuzzy logic may be inside—
•your washing machine, if your machine determines the best washing cycle from your input of water level, load size, fabric , and dirt and stain levels. After a wash cycle. The machine will repeat the cycle if the water is not clean enough.
• your vacuum cleaner, if it can automatically adjust suction based on information gathered through its infrared sensors.

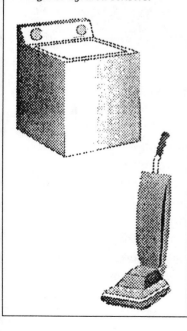

term "fuzzy" logic was coined in the 1960s by Berkeley professor Lofti Zadeh, but paradoxes can probably be blamed for the birth of this new logic. Over the centuries paradoxes have created glitches in traditional logic. True or false logic could not be used to explain such paradoxes as Eublides' *pile of sand paradox* [1] or Bertrand Russell's *class membership paradox.*[2] These paradoxes

The Mitsubishi HSR-IV is designed to avoid accidents by utilizing a neural fuzzy inference system.

go hand-in-hand with so many real life situations which cannot be answered by a simple yes or no. Traditional logic does not have room for gradations of a situation. Although fuzzy logic had its origins in the United States, it has been embraced by Eastern philosophy and scientists. Just as mathematical objects do not precisely describe things in our world, so traditional logic cannot be perfectly applied to the real world and real world situations. Traditional logic and computer programming rely on statements being either true or false (electricity being "on" or "off" — "1" or "0" of the binary system). To humanize computers, fuzzy logic must come into the picture. Fuzzy logic tries to imitate how the human brain would work. In other words, it attempts to make artificial intelligence real intelligence.   Japanese and Korean companies are leading the way with innovations that make use of *fuzzy logic*.   Among other things, they are producing computers, air conditioners, cameras, dishwashers, automobile parts, televisions and washing machines, which are enhanced with fuzzy logic technology. In the October 1993 Tokyo Motor Show, the computer in Mitshubishi's prototype HSR-IV uses fuzzy logic to imitate the information processing of the driver's brain. The computer studies the driver's normal driving habits, and then selects a response for various situations that might occur. For example, if built-in radar detects an obstacle, the fuzzy logic system decides if the driver is aware of it or not (based on past driving patterns). If the driver does not respond as predicted, it can then automatically control the brakes to avoid a collision.

Up until now many Western scientists have been reticent to use fuzzy logic, feeling that it threatened the integrity of scientific thought. Other scientists feel it enhances and expands the possibilities of computer programming by considering a broader scope of variables for the solution of a specific problem or the design of a particular product. In the past the National Science Foundation rejected proposals that dealt with fuzzy logic, but perhaps future projects will be considered in another light, especially since more United States companies, such as Ford, Otis Elevator, Pacific Gas & Electric, Motorola, and General Electric are expressing interests in its uses.

[1] 4th century B.C. Greek philosopher Eublides considered how many grains of sand make a pile.

[2] Russell's paradox —Bertrand Russell's (1872-1970) paradox deals with the idea of membership of a set. A set is either a member of itself or not a member of itself. Refer to a set which does not contain itself as a member as regular.   For example, the set of people does not include itself as a member, since it is not a person. Refer to a set which does contain itself as a member as irregular.   An example is the set of sets with more than say five elements. Is the set of all regular sets regular or irregular? If it is regular, it cannot contain itself. But it is the set of all regular sets, thus it must contain all regular sets, namely itself. If it contains itself, it is irregular. If it is irregular, it contains itself as a member, but it is supposed to contain only regular sets.

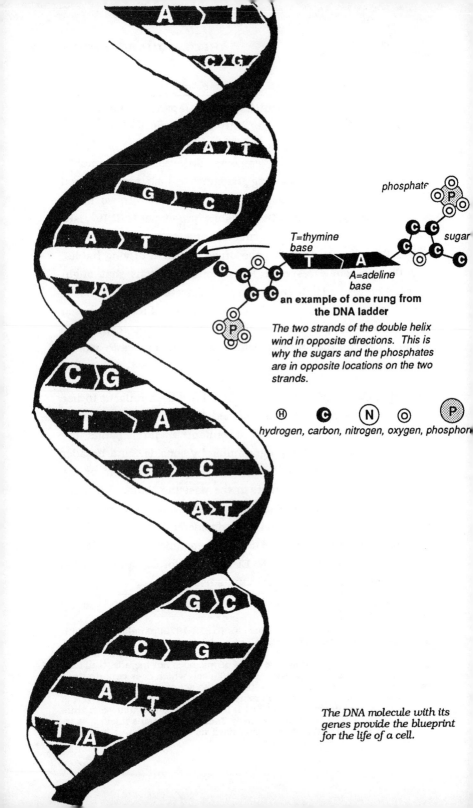

phosphate

sugar

T=thymine base

A=adeline base

an example of one rung from the DNA ladder

The two strands of the double helix wind in opposite directions. This is why the sugars and the phosphates are in opposite locations on the two strands.

Ⓗ   Ⓒ   Ⓝ   ⊚   Ⓟ

hydrogen, carbon, nitrogen, oxygen, phosphorus

The DNA molecule with its genes provide the blueprint for the life of a cell.

MATHEMATICS &
THE MYSTERIES
OF LIFE

MATHEMATIZING THE HUMAN BODY

MATHEMATICAL MODELS & CHEMISTRY

MATHEMATICS & GENETIC ENGINEERING

BODY MUSIC

SECRETS OF THE RENAISSANCE MAN

KNOTS IN THE MYSTERIES OF LIFE

*It would only be possible to imagine life or beauty as being
"strictly mathematical" if we ourselves were such infinitely
capable mathematicians as to be able to formulate their
characteristics in mathematics so extremely complex that
we have never yet invented them.*
    —Theordore Andrea Cook (1867-1928)

The scientific world is constantly using mathematical ideas to try
to unravel the mysteries of life. This chapter presents a few of the
areas in which mathematics is used in the search for answers to
such questions as how our bodies work and how life began. It
seems a mystery in itself that a subject which deals with
inanimate objects and objects of our imagination could hold the
answers to these questions.

From **The Creation of Adam** by Michelangelo. Sistine Chapel, Vatican,
Rome, Italy

MATHEMATIZING THE HUMAN BODY

blood pressure: 120/80

cholesterol: 180

LDL/HDL: 179/47

triglicerites: 189

glucose: 80

body temperature: 98.7˚F

In present day medicine, we, as patients, are bombarded with numbers and ratios that analyze our health and how our bodies are functioning. Doctors have tried to pin point ranges of numbers which are normal. Numbers and mathematics seem pervasive. In fact, in our bodies the networks of our cardiovascular system, the electrical impulses our bodies use to trigger movements, the ways in which cells communicate, the design of our bones, the actual molecular structure of genes — all possess mathematical elements. Consequently, in an effort to quantify bodily functions of humans, science and medicine have resorted to numbers and other concepts of mathematics. For example, instruments have been devised to translate the body's electrical impulses to si-nusoidal curves, thereby making it feasible to compare outputs. Readings from an electrocardiogram, electromyogram, ultra-sound — display the curve's shape, amplitude, phase shift. All of this gives information to the trained technician. Numbers, ratios, and graphs are aspects of mathematics adapted to our bodies. Let's consider other mathematical concepts and how they are linked to the body.

If you think that deciphering codes, ciphers, and Mayan hieroglyphics is exciting and challenging, imagine being able to unravel molecular codes the body uses for communication.

Science has now discovered that white blood cells are linked to the brain. The mind and body communicate via a vocabulary of biochemicals. Deciphering these intercellular codes will have an astounding impact on medicine, just as our increased under-standing of genetic codes is unveiling so many ramifications in the health field. The discovery of the double helix in DNA was another mathematical phenomenon. But the helix is not the only spiral present in the human body. The equiangular spiral is present in many areas of growth — possibly because its shape does not change as it grows. Look for the equiangular spiral in the growth pattern of the hair on your head, the bones of your body, the cochlea of the inner ear, the umbilical cord, or perhaps even your fingerprints.

The physics and physical aspects of the body also lead to other mathematical ideas. The body is symmetrical, which lends itself to balance and a center of gravity. Along with enabling balance, the three curves of the spine are very important in fitness and the body's physical ability to lift its own weight and other burdens. Artists, such as Leonardo da Vinci and Albrecht Dürer, tried to illustrate the body's conformity to various proportions and measurements such as the golden mean.

As amazing as it may sound, the chaos theory also has its place in the human body. For example, the chaos theory is being explored in the area of heart arrhythmias. The study of a heart's beats and what causes the heart of some people to beat irradically indicates that it appears to conform to ideas of chaos. In addition, the functions of the brain and brain waves and treatment of brain disorders are also linked to chaos theory.

Exploring the body on a molecular level, we find inks to mathematics. Geometric shapes, such as various polyhedra and geodesic domes, are present in the shapes and forms of various invading viruses. In the AIDS virus (HTLV-1) icosahedral

*Theodore A. Cook published this analysis of Sandro Botticelli's **The Birth of Venus**. In his book, **The Curves of Life**, he states "the line containing the figure from the top of the head to the soles of the feet is divided at the navel into exact proportions given by ...the golden section (φ) ...we have seven consecutive terms of φ in the complete composition."*

symmetry and a geodesic dome structure are found.   Knots present in DNA configurations have led scientists to use mathematical findings from knot theory to study loops and knots formed by DNA strands. The findings from knot theory and ideas from various geometries have proven invaluable in the study of genetic engineering.

Scientific research and mathematics are an essential combination to discovering the mysteries of the human body and analyzing its functions.

MATHEMATICAL MODELS & CHEMISTRY

Mathematical objects are present in many natural occurring chemical substances. Studying the molecular structure of objects, one finds pentagons in the shape of deoxyribose, tetrahedrons in sil-icates molecules, double helices in DNA, polyhedron shapes in crystals and other molecular formations. Thus it does not come as a surprise that when chemists are creating new substances they rely on mathematical models. Carbon atoms are well suited as building blocks because of the way they can be linked into chains and looped into rings or formed into 3-D molecular

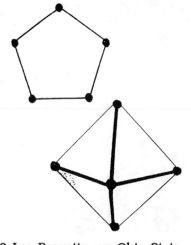

structures. For example, in 1983 Leo Paquette, an Ohio State University chemist, formed a molecule resembling a dodecahedron which consists of 20 carbon atoms surrounded by 20 hydrogen atoms. It resembled a soccer ball and was called a *dodecahedrane*. But Euclidean models are not the only ones used. In June of 1983, chemists at the University of Colorado formed a compound they named *tris* (tetrahydroxymethylethylene) which takes the shape of the Möbius strip (the single-sided single-edged discovery of August Mobius in 1848). *Tris* is composed of chains of carbon and oxygen atoms and ends in alcohol groups, which lend themselves to being clipped easily together after the chain is given a half-twist.

MATHEMATICS & GENETIC ENGINEERING

Jurassic Park has made the general public very aware of the wonders and possible horrors of genetic engineering. The drama of life unfolds within a living cell. It is not important if that living cell is human, fish, foul, plant, insect or bacteria, every living cell is a home of a DNA helical molecule. But what is this 3-D mathematical spiral doing inside a cell? What is its function? And how are genes — the transmitters the characteristics of life forms — related to DNA? Here enter the mathematical concepts of patterns, sequences, relations, one-to-one correspondence, all playing a role in unraveling the codes and mysteries of the living cell. Within a plant or animal cell,

we find a nucleus[1] in which resides chromosomes. DNA is split into strands called chromosomes, and genes are found on DNA molecules. RNA molecules and proteins are also in the nucleus along with other minuscule materials and water. Different species have a different number of types of chromosomes. For example, humans have 46, a certain flower species has 4, while a hermit crab 254 [2]. It is amazing to consider that every living cell is composed of the same six elements — carbon, hydrogen, oxygen, phosphorus, nitrogen, and sulfur. In the cell these atoms combine to form molecules of water, phosphate, and sugar. In addition, they form macromolecules (enormous molecules made up of thousands of atoms strung together) such as lipids, starch, cellulose and the even more complex nucleic acids and proteins. The DNA molecule with its genes provides the blueprint for the life of a cell. Even though every cell's essential building elements are the same, different organisms have dramatically different genetic blueprints. And although genes use the same "symbols" (elements to form the cell's code), the different structures of living things are formed by involving different combinations of sequences of these "symbols". In every cell the codes are deciphered by the same type of mechanism used to replicate cells and create proteins and other cellular structures.

The formation of genetic codes lies in understanding nucleic acids. Nucleic acids are formed from nucleotides — made from a sugar, a phosphate, and a base (there are five kinds of bases which are referred to by the symbols A, C, G, T, U)[3]. In DNA, only the four bases A, C, G and T appear. Each base can pair with only one

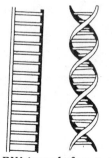

DNA is made from two helical strands that are joined at their bases to form a double helix.

other complementary base — A with T and C with G— and in DNA two strings of nucleic acids join together at their bases and form the famous double helix. It is the pentagonal shape of deoxyribose which makes the shape of the spiral and requires 10 rungs to complete a turn. On the other hand, RNA has only the bases A, C, G, and U and is usually a single rather than double strand and much shorter than DNA. The DNA diagram illustrates how the string of bases forms a specific code. When gene copying (or DNA replication) takes place the DNA double helix is unwound at breakneck speeds of over 8000rpm, and splits along the bases separating into two strands. While this is happening unattached free nucleotides in the cell with the proper base "symbol" attaches their bases to the

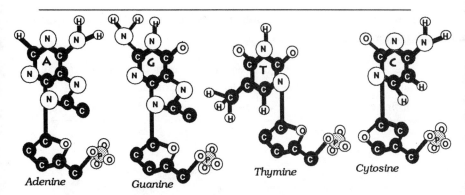

Adenine   Guanine   Thymine   Cytosine

Enormous numbers of nucleotides reside in a living cell. There are four types of nucleotides which are made up of a base and part of the helix. Note the sugar phosphate molecules are identical and can therefore fit together in any order and to form one strand of the DNA helix.

DNA's bases along each split strand. In this way two identical DNA double helices are formed by splitting the original in half and adding free nucleotides to the halves. In addition to DNA replication, genes also formulate proteins[4] and other cell functions. Genes are triggered into action by various stimuli (e.g. exposure to a particular hormone). How does a gene orchestrate its activities? The thousands of genes alined along the DNA molecule are analogous to the "on" and "off" switches of a computer's binary instructions — genetic programs or algorithms. A stimulated gene uses a chemical reaction to trip the switches of other genes of the DNA molecule. Its signal runs along the DNA molecule turning genes "on" or "off". Those turned off stop transmitting a signal, while those turned on start transmitting their own instructions. Imagine all these processes taking place simultaneously. Genetic engineering is the process of manipulating the structure of DNA molecules. These engineers have discovered ways to replicate, alter, splice, and attach code sequences to the DNA molecule, in order words change the blueprint of a cell. They can make what is called recombinant DNA, DNA that consists of a combination of bases that are taken from totally different organism, e.g. from a human cell and a bacteria cell.

Besides the study of sequences, codes and helices, mathematical shapes and structures are also important in categorizing the infinite combination of atoms and molecules that form in living

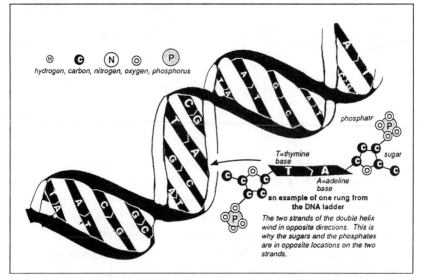

hydrogen, carbon, nitrogen, oxygen, phosphorus

phosphate

T=thymine base

sugar

A=adeline base

an example of one rung from the DNA ladder

The two strands of the double helix wind in opposite directions. This is why the sugars and the phosphates are in opposite locations on the two strands.

cells. Since these 3-D structures, such as proteins are not rigid but often appear to be knotted—the mathematics of knot theory and computer modeling are playing an ever increasing role in genetic engineering. Chaos theory is another important field here, since simple minuscule changes in the gene stimuli could result in enormous difference in the outcome, e.g. a mutation. And what about fractals? Cannot the life of a cell be considered a super sophisticated fractal-like creation where the given objects are the six basic elements of life and its rule the genetic program of the DNA molecule?

With genetic engineering techniques being developed and discovered at a frantic pace, biotech companies are racing to use relatively new science to discover new cures and to manufacture miracle drugs.   Food manufactures are seeking to use genetic techniques to manipulate genes in various produce[5] to enhance shelf life, size, taste and resistance to pests. The altering of genetic codes can now be done in a minute fraction of the time it has taken the process of natural selection to take place. The questions remain. Are these changes desirable,   effective,   damaging? Will the uninformed consumer be at risk from po-

Genetically engineered tomatoes can ripen on the vine and are bruise and frost resistant.

tential hidden allergens added through gene splicing? The responsibility that genetic engineers must bear is enormous. They are experimenting with life as it has evolved over millions of years!

[1] Protocyotes cells such as bacterial organism's cells do not have nuclei, while eucaryotes cells which are those found in plants and animals each have a nucleus.

[2] The varying numbers are influenced by many factors including the various environmental stages that species experiences during it life.

[3] A is adekine, C is cystosine, G is guanine, T is thymine, U is uracil.

[4] Like nucleic acids, proteins are also long chains of smaller units called amino acids (thus far 20 types of amino acids have been identified). Every protein is identified by a specific sequence and number of amino acids.   Enzyme proteins are responsible for directing the chemical reaction of a living cell. Where does RNA enter the picture? When the DNA is splitting during its replication process, a messenger RNA is constructed along the DNA strand matching complement bases on the DNA strand. (This process is called transcription.) The messenger RNA carries this genetic code from the DNA to the translators of the genetic code called transfer RNA molecules. Here a string of amino acids are connected in the translated sequence by ribosomal RNA and the protein is formed. Many steps and parts have been eliminated for simplicity, but the wonder of this process taking place simultaneously in living cell is mind boggling.

[5] These include: tomatoes with flounder genes to reduce frost damage, ripen on the vine and not easily bruise in shipment; potatoes with silkworm genes to increase disease resistance; corn with firefly genes to decrease insect damage. Consumer action groups, such as Pure Foods Campaign based in Washington D.C. have organized to monitor genetically engineered foods.

BODY MUSIC

We have heard of body language created by the various messages we give off by the way we move or posture our bodies. But what's body music? Some scientists[1] have been exploring patterns found in DNA from another totally different point of view— the music of genes. Two helical strands are joined at their compatible bases[2] to form the double helix chain-like formation of DNA. The bases hold

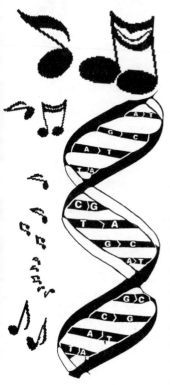

genetic codes and along with the DNA are the blueprints for the living cell. The patterns of these bases are being intensely studied in hopes of deciphering their meanings. Repetitive sequences of bases have been observed to recur throughout the gene. A musical link becomes apparent, when one considers the recurring sequences as recurring melodies of a song. In fact, many of them have been put to music, both in an octave and other intervals.

As mentioned earlier, music and math have been linked over the centuries. The Pythagoreans linked numbers and musical scales. Even Johannes Kepler (1571-1630) linked the velocities of planets in their elliptical orbits with musical harmony, although this connection today is no given scientific significance. Therefore, it is not unusual for scientists and mathematicians to seek out the music of the body.

---

[1]Dr. Susumo Ohno of the Department of Theoretical Biology at the Beckman Research Institute of the City of Hope, Duarte, California.

[2]Only certain bases pair up with each other. These are A with T and G with C. The way they appear on the chain, forms a DNA's genetic code.

This famous drawing by Leonardo da Vinci appeared in the book, *De Divina Proportione*, which Leonardo illustrated for math-

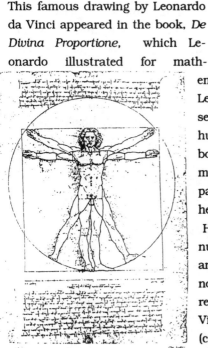

ematician Luca Paoli in 1509. Leonardo wrote an extensive section on the proportions of the human body in one of his notebooks. He determined measurements and proportions for all parts of the body, including the head, eyes, ears, hands and feet.

His proportions were based on numerous studies, observations and measurements. In his notebook, he also made reference to the works of Vitruvius, the Roman architect (circa 30 B.C.) who also dealt with the proportions of the human body. Leonardo writes of how he was influenced by Vitruvius:

> *Vitruvius, the architect, says in his works on architecture that the measurements of the human body are distributed by Nature as follows: ...If you open your legs so much as to decease your height by 1/14 and spread and raise your arms till your middle fingers touch the level of the top of your head you must know that the center of the outspread limbs will lie in the navel and the space between the legs will be an equilateral triangle.*

Leonardo adds, *The length of a man's outspread arms is equal to his height.* [1]

[1]Richter, Jean Paul, editor. *The Notebooks of Leonardo da Vinci, vol. 1.* Dover Publications, 1970, New York.

KNOTS IN THE MYSTERIES OF LIFE

From the time Alexander the Great cut the Gordian knot, knots and their many shapes and forms have pervaded many facets of our lives. Magicians, artists, and philosophers have been intrigued by such knots as the trefoil knot, which has no beginning or end. Today scientists are looking to knots as possibly holding some of the keys to the mysteries of life. These theories range from the cosmic to the microscopic worlds.

At first glance one would think there was nothing special about knots, other than keeping things secured, such as shoes or rigging on a boat. But in mathematics there exists an entire field called knot theory, and discoveries are constantly being made that link this theory directly to the physical world.

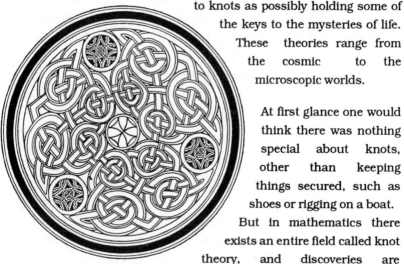

Celtic knot design from Gospel of St. John. Book of Durrow. From **Celtic Design** *by Ed. Sibbett, Jr. Dover Publications, 1979.*

Knot theory is a very recent field of topology. Its origins can be linked to the 19th century and Lord Kelvin's idea that atoms were knotted vortices that existed in ether, which was believed to be an invisible fluid that filled space. He felt he could classify these knots and arrange a periodic table of chemical elements. Although his theory was not true, the mathematical study of knots is a very current topic today.

What distinguishes mathematical knots from the everyday knots one ties, is that they have no ends. They are a closed type of loop, which cannot be formed into a circle. What mathematicians have been trying to do is classify knots and be able to distinguish different knots. Some of the important ideas that have been formulated thus far are:

- *a knot cannot exist in more than three dimensions*

- *the simplest possible knot is the trefoil knot, which has 3 crossings. It comes in a left and right handed version which are mirror images of one another*

the trefoil knot

- *there exists only one knot with 4 crossings*

- *there are only two types of knots with 5 crossings*

- *over 12,000 knots have thus far been identified with 13 or less crossings, not counting mirror images*

From left to right: the first knot has 4 crossings, the second has 5, the third has 6, the fourth and fifth have 7.

- *Consider the diagram. below. These knots are mirror opposites of each other. You would think that since they are opposites, they would undo each other when brought together. Try it! (They simply pass through one another and remain unchanged.)*

- *Now look at the
  Chefalo knot
  or false knot.
  What happens when
  the ends are pulled?
  (The knot falls apart.)*

Originally, actual models of knots were created in order to study them. To test if two knots were equivalent, one of the models had to be manipulated by hand and an attempt was made to transform it to the other's shape. If it was transformed they were identified as equivalent; if not, no conclusion was possible.

The study of knots in the field of topology tries to explain these various properties. Computers have even come into the picture. The *Geometry Supercomputer Project* [1] is using advanced computer technology to study and produce visual three dimensional representations of mathematical forms and equations, such as torus knots and fractals.

Mathematicians have developed other methods for classifying and testing knots.[2] They now look at the two-dimensional shadow it casts and write an equation describing it.[3] Many recent inroads have been made and some exciting connections discovered linking knot theory to work in molecular biology and physics. Scientists in these fields have used the findings of mathematicians and applied the new techniques in knot theory to their study of DNA configuration. Researchers have discovered that DNA strands can form loops which sometimes are knotted. Now scientists can use findings from knot theory to decide if a DNA strand they are viewing has appeared before in another knot form. They can also determine a sequence of steps in which DNA strands can be transformed to produce a particular configuration and to predict unobserved configurations of DNA. All of these findings may prove very helpful in genetic engineering.

Similarly, in physics knot theory proves very helpful when studying the interaction of particles that resemble knots. A knot configuration can be used to describe different interactions that can take place. Physicists exploring *The Theory of Everything*(TOE) believe that knots played a vital role in the creation of the universe. In their incessant search of TOE, scientists are trying to develop a mathematical model that would unify the forces of nature (electromagnetism, gravity, the strong force and the weak force). Physicists certainly believe a TOE can be found. Some have devoted their life's work in search of answers.   In today's most current TOE, some physicists believe that the essence of the universe is tied into objects called *superstrings*. A version of this particular TOE contends that the universe, all forms of matter and energy, from the instant of the Big Bang resulted from the actions and interactions of superstrings. The *superstring theory* describes the universe as 10-dimensional (9 spatial dimensions and 1 dimension of time) with its building blocks of matter and energy being these infinitesimal strings.   The theory further speculates that at the moment of the Big Bang the 9-dimensions were equal. Then, as the universe expanded only 3 of the spatial dimensions expanded with it. The other 6 dimensions remained entwined and encased in compact geometries that measure only 10^{-33} centimeters across (i.e. 10^{33} laid end to end measure 1 centimeter). Hence, these strings are believed to possess 6 dimensions within them.   Scientists are now trying to describe them by using 6-dimensional topological models. It is believed that these superstrings may either be open or closed in a loop, and make up the different forms of matter and energy of the universe by altering their vibrations and rotations. In other words, superstrings are distinguished among each other by how they vibrate and rotate.

The importance of the superstring theory is being compared to Einstein's Theory of General Relativity.   Four dimensions were hard enough to envision.   Einstein posed that length, width, height and time were needed to describe the location of an object in the universe. Ten dimensions seem out of the question. But if one thinks of dimensions as descriptive numbers that pinpoint the location and characteristics of an object in the universe, they become more comprehensible.

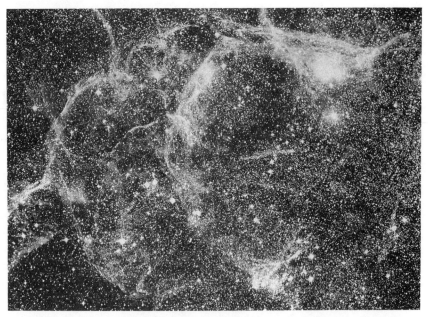

These ideas of TOE have been evolving for over 20 years. Gravity had been the wrench in the gears because the computations (involving gravity) needed to support the various forms of this theory produced mathematical infinities[4]. The breakthrough came in 1974, when John Schartz and Joel Scherks considered gravity as a curvaceous piece of geometry in the 10th-dimension similar to how Einstein described gravity in the geometry of the 4th-dimension.

The mathematics behind the superstring TOE is very high powered. The results have been most convincing. Schwartz and

Michael Green were two of the main initiators of this theory. They had been working on it for over a decade in spite of little encouragement or support from colleagues, who found the 10-dimensional world hard to accept. Their published paper finally caused physicists to take the idea seriously. Some scientists contend physicists had been avoiding ("wasting" time) researching the idea because the mathematics is too difficult. Super symmetry, six-dimensional topological models, a ten-dimensional universe, infinitesimal strings are some of the concepts needed to describe and validate the theory. As a result, this theory has turned some physicists into mathematicians and vice versa. These are just the beginning discoveries and applications of an emerging mathematical field— knot theory.

---

[1] An international group of mathematicians and scientists who work on the same supercomputer via telecommunications network with pure mathematics to solve challenging problems of geometry. *Science News*, vol. 133, p.12, January 2, 1988 issue.

[2] A "knot" that can be transformed to no twists or crossings i. e. into a loop, is a circle or unknot.

[3] The first such equations were done by John Alexander in 1928. In the 1980's Vaughan Jones made additional discoveries on the equations for knots. *Science News*, vol. 133, p.329, May 21, 1988 issue.

[4] Mathematical infinities can be caused by such operations as dividing a number by zero and raising zero to the zero power.

The Oracle office building. Redwood City, CA.

MATHEMATICS &
ARCHITECTURE

**BUCKMINSTER FULLER, GEODESIC DOMES &
THE BUCKYBALL**

**21ST CENTURY ARCHITECTURE
—SPACEFILLING SOLIDS**

THE ARCH—CURVY MATHEMATICS

**ARCHITECTURE &
HYPERBOLIC PARABOLOIDS**

**THE DESTRUCTION OF THE BOX &
FRANK LLOYD WRIGHT**

"Mechanics is the paradise of mathematical science because here we come to the fruits of mathematics."
—Leonardo da Vinci

For thousands of years mathematics has been an invaluable tool for design and construction. It has been a resource for architectural design and also the means by which an architect could eliminate the trial and error techniques of building. As comprehensive as it may seem, this is but a partial list of some of the mathematical concepts which have been used in architecture over the centuries:

Temple of Kukulkan. Chichen Itza, Yucatan.

A pyramid theme is carried out in the design of this modern office building in Foster City, Caliiornia

- pyramids
- prisms
- golden rectangles
- optical illusions
- cubes
- polyhedra
- geodesic domes
- triangles
- Pythagorean theorem
- squares, rectangles,
- parallelograms
- circles, semicircles
- spheres, hemispheres
- polygons
- angles
- symmetry
- parabolic curves
- catenary curves

- •hyperbolic paraboloids
- •proportion
- •arcs
- •center of gravity
- •spirals
- •helices
- •ellipses
- •tessellations
- •perspective

The design of a structure is influenced by its surroundings, by the availability and type of materials, and by the imagination and resource upon which the architect can draw.

St. Sophia. Istanbul, Turkey.

Some historical examples are—

- The task of computing size, shape, number and arrangement of stones for the construction of the pyramids of Egypt, Mexico and the Yucatan which relied on knowledge of right triangles, squares, Pythagorean theorem, volume and estimation.

- The regularity of the design of Machu Picchu would not have been possible without geometric plans.

- The construction of the Parthenon relied on the use of the golden rectangle, optical illusions, precision measurements and knowledge of proportion to cut column modules to exact specifications (always making the diameter 1/3 of the height of the module).

- The geometric exactness of layout and location for the ancient theater at Epidaurus was specifically calculated to enhance acoustics and maximize the audience's range of view.

- The innovative use of circles, semicircles, hemispheres and arches became the main mathematical ideas introduced and perfected by Roman architects.

- Architects of the Byzantine period elegantly incorporated the concepts of squares, circles, cubes and hemispheres with arches as used in the St. Sophia church in Constantinople.

The Roman Coloseum. Rome, Italy.

- Architects of Gothic cathedrals used mathematics to determine the center of gravity

Each of the floorplans of the three levels of this house is designed from two overlapping equilateral triangles. The triangle motif is carried out throughout the interior supports and windows.

to form an adjustable geometric design of vaulted ceilings meeting at a point that directed the massive weight of the stone structure back to the ground rather than horizontally.

• The stone structures of the Renaissance showed a refinement of symmetry that relied on light and dark and solids and voids.

With the discovery of new building materials, new mathematical ideas were adapted and used to maximize the potential of these materials. Using the wide range of available building materials —

stone, wood, brick, concrete, iron, steel, glass, synthetic materials such as plastic, reinforced concrete, prepounded concrete — architects have been able to design virtually any shape.   In modern times we have witnessed the formation of the hyperbolic paraboloid (St. Mary's Cathedral in San Francisco), the geodesic structures of Buckminster Fuller, the module designs of Paolo Soleri, the parabolic airplane hanger, solid synthetic structures mimicking the tents of the nomads, catenary curve cables supporting the Olympic Sports Hall in Tokyo, and even an octagonal home with an elliptical dome ceiling.

This tent-like structure illustrates the use of new materials and methods of constructions. Fashion Island, Foster City, California.

Architecture is an evolving field. Architects study, refine, enhance, reuse ideas from the past as well as create new ones. In the final analysis, an architect is free to imagine any design so long as the mathematics and materials exist to support the structure.

BUCKMINSTER FULLER, GEODESIC DOMES & BUCKYBALLS

Richard Buckminster Fuller was an inventor, a designer, an engineer, an author. He was an architect of ideas, a man whose visions were often ahead of their time. Among his ideas and inventions we find the following creations for which he is best known.

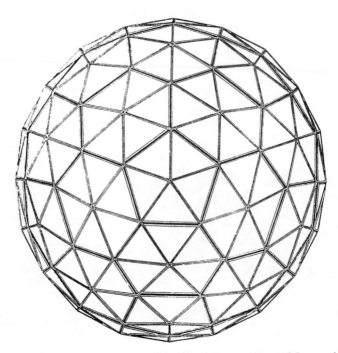

— the *dymaxion car* designed with a rear engine and front wheel drive

— the *Dymaxion House* and *Wichita House*, precursors of pre-fabricated housing for mass production, whose construction minimized use of materials, and maximized space. They were designed to be completely transportable living units.

- in the field of cartography he created the *World Energy Map* of 1940, the *Life* magazine *World Strategy Map* of 1943, and the *Dymaxion Air Ocean Map* of 1954
- the geodesic dome (triangulated space enclosing dome)
- tensegrity

The geodesic dome was Fuller's greatest commercial success, and the one with which his name has become synonymous. In his patent application he described the geodesic dome as follows:*"My invention relates to a framework for enclosing space. A good index to the performance of any building frame is the structural weight required to shelter a square foot of floor from the weather. In conventional wall and roof designs the figure is often 2500 kg per square meter. I have discovered how to do the job at around 4 kg per square meter by constructing a frame with a skin of plastic material."*[1] Buckminster Fuller's most important insight was that he saw a connection between the traditional polyhedra[2], the sphere, and architecture. This connection materialized into the geodesic dome.   Let's reconstruct a possible scenario of the evolution of his geodesic domes. Beginning with the icosahedron, he subdivided its faces into equilateral triangles. Now he circumscribed this with a sphere and projected its vertices onto the sphere. The equilateral triangles do not remain congruent. Suppose he truncated the surface of this new solid. Now the geodesic structure's shape more closely approaches the shape of the sphere, and thus also the properties of the sphere. A sphere has the least surface area for a given volume. Consequently the geodesic dome encloses more space with less expenditure of materials than conventional architectural forms. In addition, a new type of stability results. As in the soap bubble, surface tension pulls inward against the outward compression of the enclosed air— an equilibrium is reached between tension and compression. The design of traditional architecture requires the consideration of weight and support. Gravity plays a prominent role. In the geodesic structure, gravity's role is nearly irrelevant.

There was one catch, however. The geodesic domes were not spheres, so their sizes were somewhat limited. But when Fuller combined the ideas of geodesic domes with tensegrity[3] (*tensional integrity*), the size of these domes could be enormous. These structures hold themselves up not with supporting columns, beams, arches or buttresses, but with tensile forces (the pulling action of the load). Fuller's invention and patent of light-weight tensegrity construction nearly eliminated size limits of these structures. To illustrate the scope of these megastructures, consider Fuller's proposed hemispherical dome with a 3 kilometer diameter that would enclose part of New York city. In 1960 he described its installation as follows, "*A fleet of sixteen of the large Sikorsky helicopters could fly all the segments into position for 1.6 kilometer high, 3 kilometer wide dome in three months at a cost of $200 million...an area of fifty blocks which includes all of upper Manhattan skyscraper city. A dome of this kind would prevent snow and rain from falling on the protected area and control the effects of sunlight and quality of the air...*"[4]

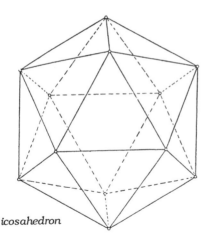

icosahedron

Fuller's geodesic dome was not without numerous setbacks in both financial aspects and acceptance. Although his gift of the Dymaxion House 'patent' to the American Institute of Architecture was rejected in 1928, his philosophical vision of providing economical housing was a driving force. During the subsequent fifty years of his career, he received honorary architectural degrees, fellowships, awards and acclaim. Equally important was the recognition brought by the over 300,000 geodesic domes that were based on Fuller's patents between 1954 until his death in 1983. Fuller's geodesic structures are not confined to the megastructure size. Geodesic structures have also been found at the molecular level. The *buckyball,* also called *buckminsterfullerene* , is a chemical polyhedron that was synthesized in the 1990s. C_{60} is a carbon molecule consisting of 60 carbon atoms located at the vertices of a truncated icosahedron.[5] The buckyball is now being synthesized in solid form for possible uses as a lubricant, catalyst, the tip of microscopy scanner, or in new battery storage cells. Scientists are also modifying the architectural structure of the buckyball (somewhat analogous to how Fuller modified

the polyhedron) to produce new theoretical molecules that would be more stable, lighter and stronger than a buckyball.[6] The conservation of materials, lightness, stability and strength of the geodesic dome are the same characteristics that interest researchers in these new molecular forms.

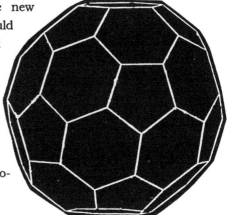

truncated icosahedron

[1] His patent was approved in June of 1954 and he received royalties on all geodesic domes that were built during the 17 years it was in effect. From *Buckminster Fuller* by Martin Pawley, Taplinger Publishing Co., New York, 1990.

[2] Polyhedra are geometric solids whose faces are polygons.

[3] In November of 1962 Fuller received a patent covering all tensile-integrity structures.

[4] From *Buckminster Fuller* by Martin Pawley, Taplinger Publishing Co., New York, 1990.

[5] It is also called a *soccerne* because of its resemblance to a soccer ball.

[6] The 168-carbon molecule called the *buckygym* — a jungle-gym-like structure — was created by researchers at the IBM Thomas J. Watson Research Center in New York.

21ST CENTURY ARCHITECTURE — SPACE FILLING SOLIDS

Over the ages the triangle, the square and the rectangle have played major roles in architectural design. Wood and stone were among the first natural materials builders used to make their shelters. Since the triangle and right angles gave the most stability from what was then known, these shapes were utilized in such structures as the pyramids of Egypt and the Yucatan. New shapes and designs evolved as knowledge, understanding, and materials improved. For example, over the centuries discoveries of the dynamics of curves and arches

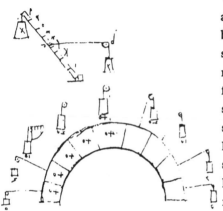

A diagram from Leonardo da Vinci's notebook, showing his work on trying to resolve the forces acting on an arch.

allowed stone and wood to be used to introduce these features in such structures as Roman aqueducts and Gothic cathedrals. With the introduction of steel, iron, glass, concrete, and bricks, new bold designs were possible. Later, plastics and synthetic materials, along with tensegritory structures, allowed architects to consider a whole new family of shapes. Our mathematical knowledge, coupled with com-

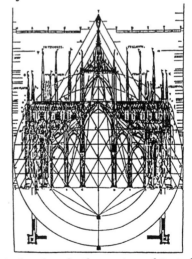

Gothic plans by the master architect of the Dome of Milan, Caesar Caesariano.

This model of Paulo Soleri's visionary project is displayed at Arcosanti in Arizona.

puter modeling, and an in-depth understanding of the various physical forces which act on a structure have greatly increased. Yet the shapes and forms of architecture remain the three dimensional objects of mathematics. Many are from Euclidean geometry, such as rectangular or square solids, pyramids, cones, spheres, cylinders. Others are more exotic forms using curved solids, tents, and geodesics. All these objects are used by the architect to both fill space and to create a living space.

What type of structures and living spaces will be designed in the 21st century? What objects can fill space? If design features involve prefabrication, adaptability and expandibility, then ideas of plane and space tessellating play prominent roles. Any shape that tessellates a plane, such as the triangle, the square, the hexagon, and other polygons can be adapted for spatial living units.

Additional units can be added to shared walls later as the need arises. Design possibilities can be very exciting with courtyards or alcoves left open for light and exterior access. On the other hand the architect may want to consider solids which pack space, the most traditional being the cube and rectangular parallelepiped. Some module designs may include rhombic dodecahedra or truncated octahedra.

Plane filling Penrose tiles & space filling truncated octahedra

The choices and options now available to architects add to today's challenges in determining which solids work best together to fill space in such a way to optimize designs, aesthetics and create comfortable livable areas. Designs and projects by such architects as Paulo Soleri, David Greene, Pier Luigi Nervi, Arata Isozaki, I. M. Pei, and others lend themselves to the use of new materials and far out geometric shapes. Now, as in the past, the feasibility of a structure is dictated by the laws of mathematics and physics — which act as both tools and measuring rods.

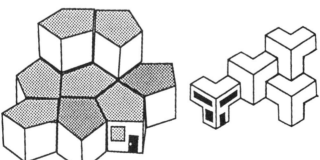

The left illustration shows how these pentagons, tessellating a plane, can be adapted to pentagonal prisms that become module space filling units. At the right is a structure formed by modifying the shape of a cube.

THE ARCH
— CURVY
MATHEMATICS

Behind the wall, the gods play;
they play with numbers, of
which the universe is made.
—Le Corbusier(1887-1965)

The arch is an elegant architectural triumph. Over the centuries the arch has taken on the shape of many mathematical curves (such as the circle, the ellipse, the parabola, the catenary) to become— *the semicircular arch, the ogee arch, the parabolic arch, the elliptical arch, the pointed or equilateral arch, segmental arch, the squinch arch, the stilted arch, the transverse arch, the horseshoe arch, the trefoil arch, triumphal arch, the relieving arch, the triangular arch, the half arch, the diaphragm arch, the corbelled or false arch.*

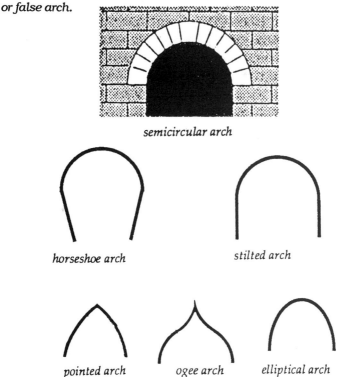

semicircular arch

horseshoe arch

stilted arch

pointed arch

ogee arch

elliptical arch

The Roman aqueduct at Segovia, Spain, constructed with 148 arches 90 feet high.

In essence, the arch is an architectural method for spanning space. The nature of the arch allows the stress to flow more evenly throughout, thereby avoiding concentration on the center. The voussoirs (wedged shaped stones), also called arch stones, form the arch's curve.   At the center is the keystone. All the stones form a locking mechanism triggered by gravity. The pull of gravity causes the sides of the arch to spread out (the thrust force). The thrust is countered by the force of the walls or of buttresses.

Until the invention and use of the arch, architectural structures relied on columns and beams, as found in Greek architecture; or stepped stones, as seen in Egyptian pyramids.   The Roman architects were the first to extensively use and develop the semicircular arch.   Add to the arch their discovery and use of concrete and brick, and an architectural revolution took place. With the use of the arch, vaults and domes, the Romans were able to eliminate horizontal beams and interior columns. The arch allowed them to relocate the weight of the structure to fewer yet more massive supports. Consequently interior space opened up. Prior to the arch, a structure was by necessity spanned with columns inside and out, and the distances

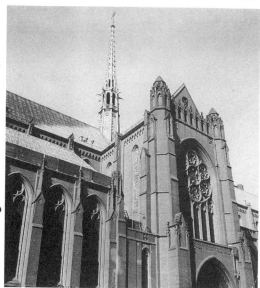

Grace Cathedral, San Francisco, California.

between the columns had to be carefully calculated so as not to have the spanning beam collapse under excessive stress.

The Roman arch was based on the shape of the circle. Over the centuries, architects began to deviate from the circle, first to the elliptical (or oval) arch, then to the pointed arch. As a result structures became higher, allowed more light to enter, and enclosed more space. The shape of the arch dictated which parts of the structure bore the weight. While the semicircular Roman arch bore the load over the span to the walls, the pointed Gothic arch had the load flow through the arch to the outside of the building's buttresses, thus making it possible for higher ceilings.

The arch is not passé. As with all architectural ideas, its concept and use are still evolving. With the invention and use of new types of building materials, architects can combine and use a multitude of mathematical curves and shapes in their creations.

ARCHITECTURE & THE HYPERBOLIC PARABOLID

Some architectural structures have been designed in less recognizable shapes. A striking example is the hyperbolic paraboloid used in the design of *St. Mary's Cathedral* in San Francisco. The Cathedral was designed by Paul A. Ryan and John Lee and engineering consultants Pier Luigi Nervi of Rome and Pietro Bellaschi of M.I.T.

At the unveiling when asked what Michelangelo would have thought of the Cathedral, Nervi replied: "He could not have thought of it. This design comes from geometric theories not then proven."

The top of the structure is a 2,135 cubic foot hyperbolic paraboloid cupola with walls rising 200 feet above the floor

St. Mary's Cathedral, San Francisco, California.

and supported by 4 massive concrete pylons which extend 94 feet into the ground. Each pylon carries a weight of nine million pounds. The walls are made from 1,680 prepounded concrete coffers involving 128 different sizes. The dimensions of the square foundation measure 255' by 255'.

A hyperbolic paraboloid combines a paraboloid (a parabola revolved about its axis of symmetry) and a three dimensional hyperbola.

hyperbolic paraboloid equation:

$$\frac{y^2}{b^2} - \frac{x^2}{a^2} = \frac{z}{c^2} \qquad a,b>0, \; c \neq 0$$

THE DESTRUCTION OF THE BOX
THE ARCHITECTURE OF FRANK LLOYD WRIGHT & THE LIBERATION OF SPACE

The work of Frank Lloyd Wright has a definite style, yet his structures are so diverse that the style does not lie in the similarities of his buildings, but rather in the philosophy that the structure projects. In his words, "Architecture is the scientific art of making structure express ideas. " His architecture has come to be called *organic architecture* — encompassing landscape, materials, methods, purpose and imagination in a special way.

Wright's Marin County Civic Center was one of his final designs.
Marin County, California

Wright's designs of structures in which internal and external space become one had a profound impact on architecture. He designed buildings so that the outside came inside. He called this

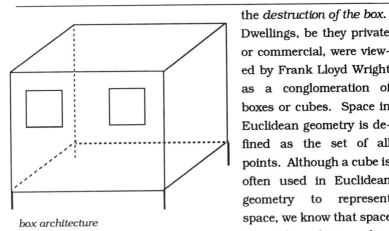

box architecture

the *destruction of the box.* Dwellings, be they private or commercial, were viewed by Frank Lloyd Wright as a conglomeration of boxes or cubes. Space in Euclidean geometry is defined as the set of all points. Although a cube is often used in Euclidean geometry to represent space, we know that space has no boundaries or limits. Wright wanted his works to give the feeling of space— the flowing of points from the inside to the outside. Thinking along these lines, he discovered a way to eliminate the traditional box from his designs. He sought a change from the feeling of confinement and separation from the outside world which characterized box architecture. Wright realized that the potential of certain building materials had not been utilized. These materials — steel and glass —

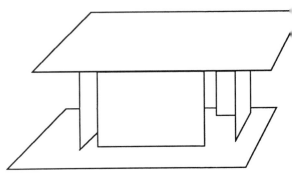

along with innovative design changes furnished the means to do away with the box, thereby allowing the merging of interior and exterior space.

Wright's designs did away with corners of the box by removing the supports from the corners and relocating them along the walls by using cantilevers. Thus, the inhabitant's eyes were not bound or led to corners. Space was allowed to flow.

By replacing post and beam construction with cantileverage supports, walls were no longer viewed as enclosing walls, but rather as
independent and unattached. Any one of these walls could be
modified by either shortening, extending or redividing.

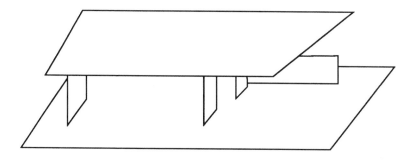

Wright did not stop with freeing the horizontal plan, but also the
vertical. He did away with cornice, and opened up the top to the
sky. His designs eliminated the stacking and duplication of boxes.
Instead, he used columns, and made them part of the ceiling,
thereby creating a continuity of form.   Now space inside and outside of a structure could move in all directions. The freedom of allowing space to move in and out of a structure is the essence of *organic architecture.*   "So organic architecture is architecture in
which you feel and see all this happen as a third dimension ...
space alive by way of the third dimension." [1]

[1]Frank Lloyd Wright, *An American Architecture*, ed. Edgar Kaufman. New
York: Bramhall House, 1955.

People have always played games and done puzzles. Books on the origins of various games give numerous examples of ancient games that are still played today. This illustration is from an Egyptian papyrus dating back to 1200 B.C.. It is a humorous rendition of a goat and a lion playing the game of senet. Senet was one of the most popular games of its time, and was played by all sectors of society. Unfortunately, there is no remaining documentation on exactly how it was played, but a possible version of senet has been devised using the findings and work of archeologists.

THE SPELL OF
LOGIC, RECREATION
& GAMES
THE THREE MATHEMATICAL MUSKETEERS

> *Logic is the art of going wrong with confidence."*
> **—Morris Kline**

Logic and mathematics definitely go hand in hand. But most people do not consider mathematics fun and games. Yet, games and recreations are an integral part of mathematics. The development of many mathematical ideas are the result of a person's pursuit of an intriguing notion or problem. Some people seem to be driven by an invisible force to work on problems and puzzles for hours on end. They belong to those who inherently enjoy mathematics and are fascinated by it. Before realizing it, one may have spent hours even days exploring different ramifications of what ostensibly started out to be a simple pastime. As history attests problems, challenges, games and pastimes have sometimes lead to remarkable discoveries, and even to the creation of new fields of mathematics. In fact, the famous

Circa 212 B.C. Syracuse fell to the Romans. At the time, Archimedes was working on a mathematics problem in his home. When a Roman soldier entered and ordered him to stop working, Archimedes did not pay attention. Angered, the soldier killed Archimedes with his sword.

Greek mathematician Archimedes was killed because he was absorbed in a mathematics problem. The following pages will reveal some of the games, puzzles and mental calisthenics that mathematicians love to play.

MATHEM...
MYSTERY TALES

Mathematical mysteries have been around for centuries. Some of Lewis Carroll's works would fall into this category. Today, these mysteries have entered the popular scene with such books as *The Puzzling Adventures of Dr. Ecco* by Dennis Sasha and *Visitors From The Red Planet* by Dr. Cryton.

Logic problems, as illustrated below, can also be embellished with stories.

* * *

The teacher started to read—

It was one of those week s at the circus, when everything seemed to go wrong. First, one of the horses for the acrobat dancers went lame. Next the head clown threw a fit because the fat lady's child got into his makeup. Then Madrè and his wife had an argument over the trapeze artist. The final blow, came when Madrè was found dead under the big top. Next to his body was his cane, which he occasionally used. An overturned glass of water was on his desk, and a minute pile of sawdust was near his body.

The teacher stopped reading, and asked, "Well, what do you think happened to Madrè ? How did he die? You can ask me any question I can answer with a 'yes' or 'no'. So put your on thinking caps, and see where your logic leads you". Hands began to go up immediately. These logic stories always seemed to spark even the shyest student. The questions started and the drama of the circus death began to unfold.

A student asked,"What did Madrè do at the circus?" "Remember, only yes or no questions," the teacher cautioned.

"Was Madrè a manager?" Carol asked. "No". "Was there any sign of violence?" "No," the teacher replied.

"Was there anything else near the glass?" "Yes, there was." "Is it important to know what it was?" "Yes"

"Was it a sandwich?" Bill blurted out. "No."

"Was it a pill?" Tom asked. "Yes."

"He died because he had not taken his medicine?" Tom continued. "No, not exactly," the teacher's replied.

"Was there a pencil sharpener on his desk?" Terri asked. "No"

"We have to find out how he died," Terri told her fellow students. The students began discussing the situation amongst themselves, before continuing the questions.

"We know he was not well, since he was taking medicine. And we know he was not killed by a violent act," Tom said, analyzing the death scene.

"Right. He must have died from his illness," Gary said. At which point he directed a question to the teacher, "Did he die from a heart attack?" "Yes."

"Well, that was an easy one to solve this time." Gary said, feeling a bit smug. "We're not done, Gary," Barbara pointed out. "We haven't found out what caused his heart attack."

"Sure we have, he didn't take his medicine. Right Mr. Mason?"

Gary asked. "Wrong! Gary," Mr. Mason replied.

The students went back into conference. "Have we considered all
the initial facts of the case?" Terri asked her colleagues.

"No," Bob, who rarely talked in class, replied. "We haven't
considered why the sawdust information was mentioned, what his
job was at the circus, and what about the cane, and Madrè's wife
and the trapeze artist."

"You're right, Bob. We need to address all these," Terri said
excitedly. "The cane, the sawdust, the heart attack, and the
marital problems." Tom muttered thinking out loud. "I've got it!
The cane and the sawdust go together. For some reason the cane
was shortened with a saw, thus leaving the sawdust." — "Yes," the
teacher replied. Tom continued, " Perhaps Madrè got upset
because someone wrecked his cane." "NO"

"Perhaps he didn't know his cane was shortened?" Bob asked. "That's right," Mr. Mason replied. "Then, when he stood with it he was very upset, but why?" Bob wondered. "Because it didn't fit

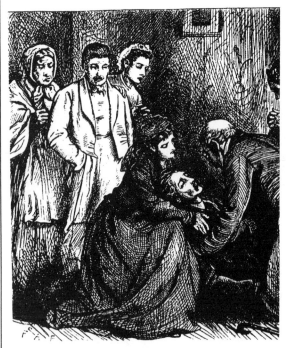

him. He was too tall for it!" Carol shouted. The excitement that the solution was close at hand spread among the students. "So far, so good," the teacher said, encouraging them to continue. "But why would thinking he had grown taller upset him so?" Gary asked the other students. "This time I've really got it," Tom yelled, waving his hand in the air and finding it difficult to contain his enthusiasm. "Madrè was a midget!" The student's excitement was contagious. The teacher shouted, "YES! What else?"

This time Terri's hand shot up in the air. The teacher nodded to her, and she began to summarize the scenario— "Madrè was a midget. His wife and the trapeze artist sawed off some of his cane, one day when he wasn't using it. When he went to use his cane, it was too short for him. Madrè thought he was growing taller, which aggravated his precarious heart problem. He couldn't get to his medicine in time. How does that sound Mr. Mason?"

"That's it! You've all done a great job on this one! Maybe I'll stump you with one next week." The teacher replied as a smile crossed his face.

Here are a few logic stumpers to try out with a friend or group of friends.

*   *   *

(1) It was Tuesday and Tom and Jerry were at the same job. When it came time for Tom to go home, Jerry wouldn't let him go home. WHY?

(2) Eric walked into a bar, and asked the bartender for a glass of water. The bartender looked at Eric for a moment, then pulled out a gun and pointed it at Eric. Eric was startled for a moment, and then said thank you and walked out of the bar without having drunk the glass of water. WHAT IS THE SCENARIO?

(3) When Mary came home she walked into the kitchen. She suddenly let out a scream when she discovered her dead husband on the floor. Along side was water from a bowl that had been on the table and was now tipped over on the floor. The window over the kitchen table was ajar. WHAT HAD HAPPENED?

PUTTING YOUR LOGIC TO WORK

RECTANGULAR VOLUME ILLUSION

Our eyes begin to jump with this puzzle. Determine how many 2x1x1 blocks make up this structure.

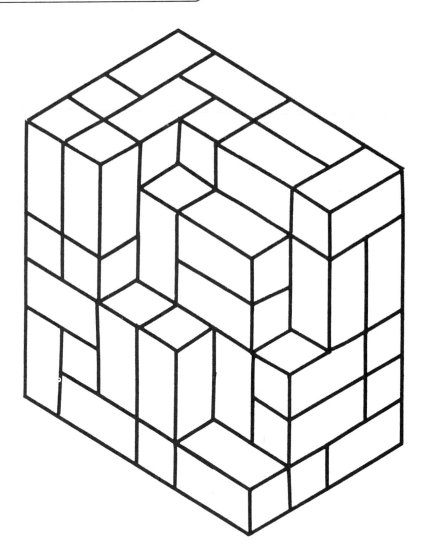

THE IMPOSSIBLE POSSIBLE

How is it possible to remove the string with two buttons from the paper trap without removing the buttons from the string, cutting the string, or tearing the paper?

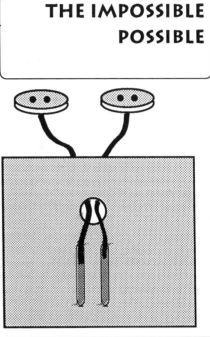

STACKING THE DIE

Assume you could walk around this stack of dice and see all exposed faces. Find the sum of the pips on the hidden faces of the dice.

ARE ALL WHEELS ROUND ?

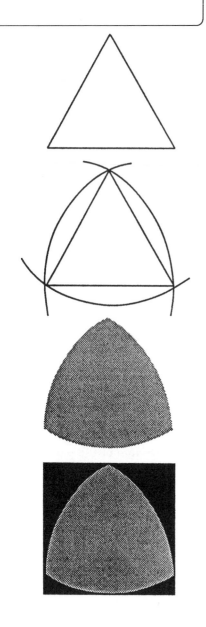

A rouleau is a fascinating object. To construct a rouleau, start with an equilateral triangle. From each vertex, set the compass to the radius of the side of the triangle. Make an arc through the two opposite vertices. If you make the rouleu out of cardboard, you can test how well it rolls. In fact, it's as smooth as if it were a circle. The center of the rouleau is the intersection of medians of the equilateral triangle. You can stick a pencil through this center and use the pencil as an axle. The rouleau can also be inscribed in a square, and when placed in a circumscribing square, it can rotate within the square. As a result, it has been used in the Wankel engine, in which the rouleau-shaped piston rotates inside a square shaped enclosure, as in Mazda automobile engines.

THE BLOCK LETTER PUZZLE

Cut out these pieces, or rearrange them in your head. Some may need to be rotated, flipped or translated. They form a capital letter of the English alphabet. Good luck!

HOW A PUZZLE CAN BE A TURNING POINT

Simeon Poisson (1781-1840) had difficulty trying to find a career that suited him. His family urged him to pursue medicine or law, but he seemed to lack the talent and/or desire. Apparently a puzzle put Poisson on track. The story has it that on a trip someone gave him a puzzle similar to the one below, which he easily solved. Realizing he had an aptitude and interest in things mathematical, he discovered that he wanted to work with mathematics. He became well known for his work on celestial mechanics, electricity, and magnetism, and for his discoveries of applications to integrals and Fourier series. In addition, he studied probability theory, and discovered the Poisson distribution.

The milkman had two 10 quart pails of milk. Two customers want two quarts each in their containers. One has a five quart container and the other has a four quart container. How does the milkman solve the problem? This version of the problem was created by Sam Loyd.

IS IT POSSIBLE?

Using only a straightedge on which distances can be indicated, can you bisect the given angle, and prove why it is bisected?

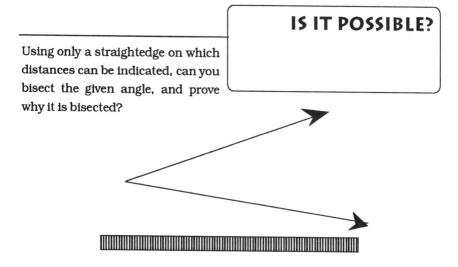

CAN YOU MAKE A LATIN SQUARE?

If no number is repeated in any row or column of a square of 18 domino tiles, it is called a Latin square. The square below is not a Latin square, since numbers repeat in some rows or columns. For example, the top row has two 6s and two 4s and the column on the right side has two blanks and two 4s. Either rearrange this, or start a new one to make a Latin square.

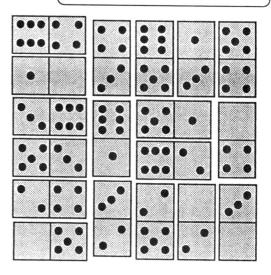

THE GAMES MATHEMATICIANS PLAY

THE CHAOS GAME

1) label three points on a sheet of paper as shown, (1,2); (5,6); (3,4).

2) Pick a starting point at random. Role a die.

3) Rule: take 1/2 the distance between the starting point and the point which has the number on the die that was just rolled.

4) Continue step three indefinitely.

(1,2)   **(5,6)**

(3,4)

It is not the first time that randomness turned out to conform to some order. 18th century French naturalist, Count Buffon, devised the *Needle Problem* and linked π to the probability of favorable outcomes. Similarly, in 1904 R. Chartres found that the probability of two numbers, written at random, being relatively prime is $6/\pi^2$.   Michael Barnsley first considered using randomness as a basis for modeling natural shapes. Hence, he invented *The Chaos Game*. There are many ways to play the game.

For example, a coin or die can be used to generate a random occurrence. The rules are also flexible. The version above uses three fixed points, a random point, and a die. The resulting design is astonishing. A chaotic process gives rise to order!

Another variation is to choose a point's location at random. Make up a rule for when a tossed coin turns up heads (e.g. move two units to the right and one up) and one for when it turns up tails. Begin tossing and see what happens.

In addition to discovering that the limit of a random process (following a set of rules) produced fractals, Barnsley also devised a random method to reproduce a given shape.

THE OLD MATHEMATICAL GAME OF RITHMOMACHIA

Rithmomachia can be a very mathematical game. When advanced levels are played, a good understanding of number theory and a strong strategy are essential. Rithmomachia dates back to the 11th century A.D., probably originating either in Byzantium or Alexandria. It was during the Middles Ages, that the intelligencia made Rithmomachia the game of choice and considered it superior to chess.

Physical set-up of game

The playing board is rectangular and is composed of 8x16 squares units. The pieces used are in the shapes of circles, squares, triangles or pseudopyramids. One opponent has white pieces called even pieces, and the other's are black pieces and referred to as odd. Each player has 8 circles, 8 triangles, 7 squares, and 1 pyramid. The white pyramid has six faces, whose numbers total 91, and is composed of two triangles (one with the number 4 and the other with 1) , two circles (with 36 and 25), and two squares (with 16 and 9). The black pyramid has five faces whose numbers total 190, and is made-up of two triangles (with numbers 36 and 25), two squares (with numbers 64 and 49), and one circle with the number 16. The shape of the piece dictates the number of spaces a piece can move. For example, the square can move 4 consecutive empty spaces in any direction, a triangle moves 3, a circle moves 1, and the pyramid can move as a circle, triangle or square depending on which face it is using. When beginning the game the pieces are set-up as illustrated in the diagram.

Object of game

The object of the game is to capture your opponent's pieces and make certain number combinations which constitute a *victory*.

**How to capture your
opponent's pieces.**
Pieces are captured
by potential moves or
by actually moving.
The methods of
capture are:

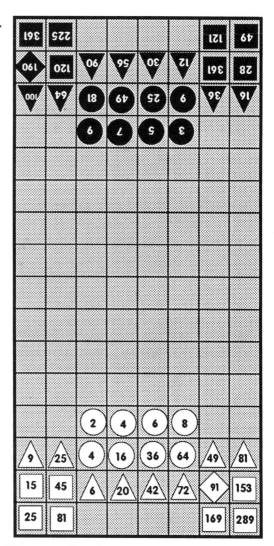

1) *Siege capture* is
done by surrounding
an opponent's piece
on all four sides.
Then that piece is
removed.

2) *Meeting capture.* If
you move a piece the
required number of
spaces and land on
an opponent's piece,
then you take that
opponent's piece
without actually
moving your piece.
For example, if white
triangle #36 moves 3
spaces and lands on
black circle#36, the
black circle #36 is

removed without actually moving the white triangle #36.

3) *Assault capture.* If a piece's number value *times* the number of
spaces in its move would land next to an opponent's piece with the
number equaling the product, then the player takes the

opponent's piece without actually doing the moving. For example, a white circle #4 removes black square #28 if 7 spaces lie between the two pieces.

4) Ambuscade. If two pieces of a player can move on either side of an opponent's piece and the sum of these two pieces equals the opponent's piece number, then the player takes the opponent's piece. For example, if black triangle #12 can be straddled by white circles #4 and #8, then black triangle #12 is removed without actually moving the pieces.

It is difficult to capture a pyramid, the number 91 and 190 must be used, unless it is done by siege or the base square (the #36 or #64) is captured when it is a base. If one of the pyramid's faces is threatened by one of the capturing methods, ransom can be collected in the form of an opponent's piece of equal or acceptable value as the threatened face.

Players decide before beginning the game, what will constitute a *victory* or *end the game.* Here are some possible ways to end the game. Some are simple, while others are quite complicated and involved.

Possible ways to end the game , i.e. gain a victory.

1) De corpore. The players agree on a number of pieces to be captured, that will declare that player the winner.

2) De bonis. Players agree on a number value target. If the sum of the number value of the pieces a player captures will total or exceed that target number, then that player is the winner.

3) De lite. Here a win depends both on the total number value of the pieces and the total number of digits of the pieces. For example, target value could be 160 with digits equaling only 6. So a player with pieces 121, 9, 30 would win, while the player having 56, 64, 28 and 15 would not (these pieces have 8 digits in all).

4) *De honore.* This win depends on the number of pieces and their number value. For example, it can be agreed that 5 pieces totaling exactly 160 will win. It is clear that the players must be familiar with the number values of their pieces and their various sums.

5) *De honore liteque.* This win requires three conditions be met— a specified number value, a specified number of pieces, and a specified number of digits of the pieces.

> *The following victories are meant for expert players and involve arithmetic, geometric, and harmonic progression. The pieces used can be from both players, but one must be your opponent's piece.*

6) *Victoria Magna.* Arrange three captured pieces in either an arithmetic, geometric, or harmonic progression. e.g arithmetic example— 2, 5, 8;   geometric example— 36, 49, 64; harmonic example— 6, 8, 12

7) *Victoria Major.* Display four pieces that can be combined to form two out of the three possible progression. For example, the pieces with value 2, 3, 4, and 8 give the arithmetic progression 2, 3, 4 and the geometric progression 2, 4, 8. Note also that 2, 4, and 8 are white pieces, while 3 is a black piece.

THE GAME OF HEX

The game of Hex was invented in 1942 by Piet Hein.

The game board for hex is made up of hexagons. There are eleven hexagons in each direction. As illustrated, the first player is not allowed to play his or her piece in any of the gray

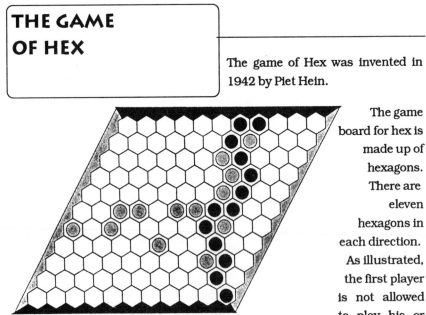

In this particular game, the player with the black pieces was the winner, since that player was able to make a path to the opposite side.

hexagons. After the first move, one can move on these hexagons. Players alternately take turns placing a piece down.

Object: Each player alternately

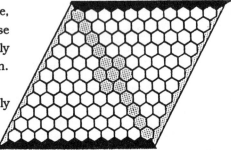

After the first move, one can move on the gray hexagons.

places his or her piece down, trying to connect a path to the opposite side. The first player to make a path to the opposite side is the winner!

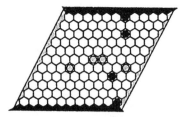

This board shows some beginning moves

SOME MATHEMATICAL RECREATIONS

Polyominoes are formed
by joining congruent squares—

monomino ☐

domino ☐☐

trominoes ☐☐☐ ⌐☐

tetrominoes ☐☐☐☐ ⊞ ⌐☐☐ ⌐☐☐ ⌐☐

pentominoes, and so on...

Polyhexes are another group of entertaining mathematical objects
which are formed from joining congruent hexagons.
Here are the configurations for up to 4-hexes (tetrahexes):

monohex ⬡   *bihex*

trihexes

tetrahexes

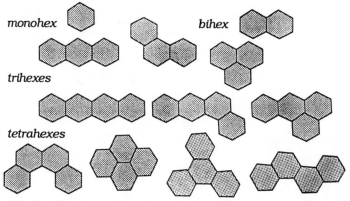

Which of the three following shapes can be tessellated using one of
each of the seven tetrahexes shown above?
Have fun!

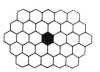

LEWIS CARROLL CHANGES "FOUR" TO "FIVE"

four
foul
fool
foot
fort
fore
fire
five

Mathematician Charles L. Dodgson, better known as Lewis Carroll, created the game of *doublets*. The English magazine *Vanity Fair* ran a number of contests in 1879 featuring Carroll's game. The game begins with two words of the same length. The object is to change the first word to the other word by forming other successive words of the same length. *You may change only one letter at a time*, trying to make the least number of transformations. Here are some others to try—

Cover EYE with LID

Prove PITY to be GOOD

Get WOOD from TREE

Change OAT to RYE

DOMINIZING PENROSE TILES

The kite and dart tiles were discovered by British mathematical physicist Roger Penrose in 1974. These two tiles tessellate a plane in a very special way. They phenomenally produce an infinite number of nonperiodic tilings (no basic pattern in the design repeats itself on a regular basis, when the eye moves up or down) of a plane. Playing a game of dominoes with them would be both a challenge and produce some very interesting designs.

Construct the tiles in the following way:
Each tile is divided by a line into two triangles, which can have one, two, three, or no dots.

Now try out a game. You may want to increase the number of dots used and see how the game proceeds.

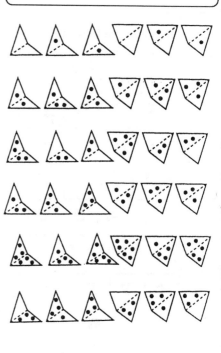

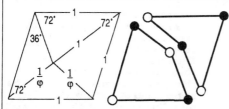

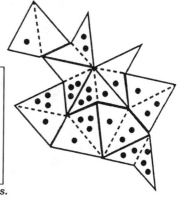

The mathematical formation of the Penrose tiles.
Note: φ is the golden mean which equals $(1+\sqrt{5})/2 \approx 1.618...$

MAKING A HEXA-TETRAFLEXAGON

Flexagons are a wonderful mathematical recreation that all ages can enjoy. In addition, those desiring more than mere recreation can delve into the mathematics of flexagons. Here is how to make a hexatetraflexagon— it has six faces and four sides.

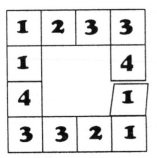

front side

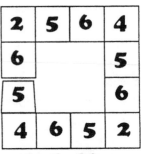

back side

step(1) Starting with the front side, fold adjacent 3's unto each other. Do the same with adjacent 1's. The resulting figure would look like —

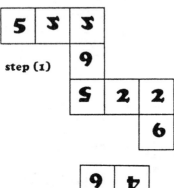

step (1)

step (2) Fold adjacent 2's unto each other. This pattern of numbers will appear on one side.

step (3) Place the two 4's on top of each other, and tape along one edge. The resulting shape is a square with 6's on one side, and 5's on the other. Join the outer edges of 6 & 5 with a piece of tape. Turn over and flex the 5's side along the center. Can you find all 6 faces?

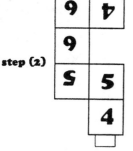

step (2)

Geometric dissection problems can
be fascinating. Problems such as
this well known rectangle dis-
section puzzle, which appears in Sam Loyd's *Cyclopedia of Puz-*

GEOMETRIC DISSECTIONS & THE CURRY TRIANGLE PUZZLE

zles,   create a paradoxical type of
situation:

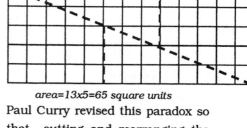

area=13x5=65 square units

Paul Curry revised this paradox so
that   cutting and rearranging the
parts resulted in an identical figure

area=8x8-64 square units

with a hole inside the figure. Here is how it works for this triangle.

Suppose
one side of
the triangle
is shaded
and the
other is
not. Cut it
apart.
Turn over
and

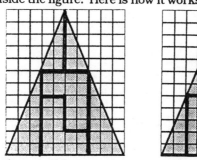

The area of the original triangle is 60 square units. The re-
vised triangle's area is 58 square units. Summing the area
of each of the pieces gives a total of 59 square units. There-
fore, the hole is 1 square unit.

reassemble the shaded parts as shown. Supposedly it is the same
size triangle, yet it has a hole in the center. What's the
explanation?

THE SQUARE TRANSFORMATION

Determine how to cut this shape into three pieces with two straight cuts, so the three pieces can be rearranged into a square.

MAGIC SQUARES & OTHER RECREATIONS

I make no question but you will readily allow the square of 16 to be the most magically magical of any magic square ever made by any magician.
                —Benjamin Franklin

PLAYING WITH MAGIC SQUARES

Imagine the formation of numbers into a square with special manipulative properties. To many people in the past these squares of numbers seemed to possess magical qualities or powers. The earliest record of a magic square appears in China around

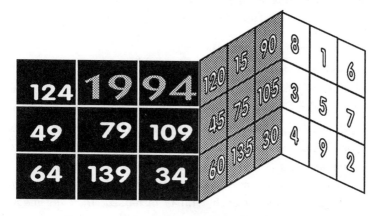

2200 B.C. when a 3x3 magic number square, called lo-shu, was seen by the Emperor Yu on the back of a divine tortoise along the banks of the Yellow river. In 9th century A.D. Egypt, ink blots were placed on a silver tray engraved with a magic square and the forms were read just as fortune tellers read tea leaves. In the 8th century A.D. certain alchemists believed that these squares held the key to converting metals to gold. Islamic magic squares often carried special meanings when the numbers of the squares were arranged to utilize their letter numerals. The Arabs used different arrangements of the 3x3 magic square to signify the signs of fire, water, earth, and air. In addition, each planet had a magic square associated with it, and magic squares were used in their work with astrology. These "mystical" squares were used in a variety ways — placing a magic square over the womb of a woman in labor, embroidering them on the garments of soldiers, placing them on the facade of a building for protection, wearing magic square amulets for luck or healing.

The magic square on the previous page was transformed into a magic square with the year 1994 on it. Starting with a 3x3 magic square, every one of its numbers were multiplied by 15. Then 4 was added to each number in the resulting square.

Over the centuries many properties and methods of construction have been discovered and developed.

Here are some of these properties—

1) Each row, column, and corner-to-corner diagonal total the same number. Each magic square has a magic constant that can be obtained in one of the following ways:

(a) A magic square's order is determined by the number of rows or columns it has. A magic square composed of the natural numbers 1, 2, 3, ..., n^2 magic constant value is $(1/2)(n(n^2+1))$, where n is the square's order. In general, if i is the initial term and β the amount each term is increased, then the magic constant $= i \cdot n + \beta(n/2)(n^2-1)$.

(b) Take any size magic square and starting from the left hand corner, place the numbers sequentially along each row. The sum of the numbers in either diagonal will be the magic constant.

2) Any two numbers (in a row, column, or diagonal) that are equidistant from its center are complements (2 numbers whose

200	217	232	249	8	25	40	57	72	89	104	121	136	153	168	185
58	39	26	7	250	231	218	199	186	167	154	135	122	103	90	71
198	219	230	251	6	27	38	59	70	91	102	123	134	155	166	187
60	37	28	5	252	229	220	197	188	165	156	133	124	101	92	69
201	216	233	248	9	24	41	56	73	88	105	120	137	152	169	184
55	42	23	10	247	234	215	202	183	170	151	138	119	106	87	74
203	214	235	246	11	22	43	54	75	86	107	118	139	150	171	182
53	44	21	12	245	236	213	204	181	172	149	140	117	108	85	76
205	212	237	244	13	20	45	52	77	84	109	116	141	148	173	180
51	46	19	14	243	238	211	206	179	174	147	142	115	110	83	78
207	210	239	242	15	18	47	50	79	82	111	114	143	146	175	178
49	48	17	16	241	240	209	208	177	176	145	144	113	112	81	80
196	221	228	253	4	29	36	61	68	93	100	125	132	157	164	189
62	35	30	3	254	227	222	195	190	163	158	131	126	99	94	67
194	223	226	255	2	31	34	63	66	95	98	127	130	159	162	191
64	33	32	1	256	225	224	193	192	161	160	129	128	97	96	65

This is Benjamin Franklin's super duper 16x16 magic square. It has all the properties of the regular magic square except its corner to corner diagonals do not total to its magic number, 2056. But as the diagram illustrates its magic number pops up in so many other ways, such as — broken 8-diagonals, broken 8-parallel rows, any 4x4 square; and perhaps you can find more. In the 1952 Journal of the Franklin Institute, Albert Chandler contends that this magic square is not Franklin's original, but one that was set incorrectly by a printer.

sum equals the sum of the largest and smallest numbers of the square).

3) There are many ways to transform an existing magic square into a new magic square.

(a) Any number may be either added or multiplied to every number of a magic square, and a new magic square will be created.

(b) If two rows or two columns, equidistant from the center are interchanged, the resulting square is a magic square.

(c) Interchange quadrants in an even order magic square.

(d) Interchange partial quadrants in an odd order magic square.

There are many methods one can use to construct magic squares. A few are—

1) **For odd order squares** — *a) the staircase method (by De la Loubère) , b) the pyramid method, c) the knight's tour method for odd order greater than 3x3, d) the Philip de la Hire method of 1705.*

2) **For even order squares** — *a) the diagonal replacement method for a 4x4; b) 8x8 diagonal method.*

3) **For any order magic square** — *The methods that do exist are complicated and involved. One such uses borders and was developed by B. Frénicle.*

Considering the lengthy history of magic squares, it is no surprise that more has been written on magic squares than any other mathematical recreation. Some people have spent hours, others days, yet others months developing methods for creating magic squares, discovering their properties, developing new ideas about them, and playing with their magic. Among the vast list we find — Benjamin Franklin's super duper 16x16 magic square— the magic square magician*, who can create any 3x3 magic square with any starting number almost instantaneously— magic lines' (segments connecting the consecutive numbers of a magic square) dazzling designs that are also used to classify the squares as symmetric or asymmetric — magic circles, cubes, hexagons, stars, spheres, magic domino squares, prime number magic squares, pan-diagonal magic squares. And the list continues to grow as more people get hooked on playing with magic squares.

*Anyone can be a magic square magician. Just memorize the basic 3x3 magic square. Challenge someone to place a number in any location of a blank 3x3 square. For example, here the number 78 was placed where the the 2 is located in the basic magic square. You immediately create a new magic square by filling in the blanks, recalling that one of the properties of a magic square is that a new one can be created by adding a constant to each number of an old magic square. In this case 76 is added to each number of the basic 3x3 magic square .

8	1	6
3	5	7
4	9	2

84	77	82
79	81	83
80	85	78

?	?	?
?	?	?
?	?	?

The role of magic squares has changed significantly over the centuries. Today they are viewed as fascinating, challenging mathematical recreations.

BENJAMIN FRANKLIN'S MAGIC CIRCLE

Benjamin Franklin was a magic square enthusiast. In fact, while he was a clerk at the Pennsylvania Assembly, he often relieved the tedium of his work by making magic circles. For his magic circle, the numbers along a radius of the largest circle all total the same amount, as do the numbers that fall in between intertwined circles.

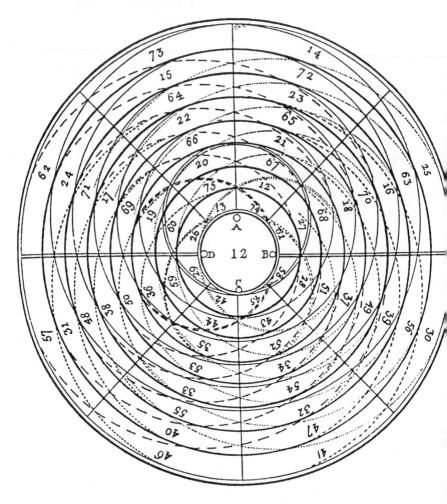

LEONHARD EULER & THE KNIGHT'S TOUR

This wonderful magic square was created by Leonhard Euler in the 18th century. As in most magic squares its rows, columns, and diagonals total the same number, in this case 260. In addition, there are four squares within it, whose rows and columns each totals 130. But even more fascinating is how a knight chess piece, starting at 1, can land on every number of the entire square sequentially from 1 through 64, by just following the moves allowed of the knight.

1	48	31	50	33	16	63	18
30	51	46	3	62	19	14	35
47	2	49	32	15	34	17	64
52	29	4	45	20	61	36	13
5	44	25	56	9	40	21	60
28	53	8	41	24	57	12	37
43	6	55	26	39	10	59	22
54	27	42	7	58	23	38	11

THE KÖNIGSBERG BRIDGE PROBLEM UPDATE

The city of Königsberg was founded by Teutonic Knights in 1308, and served as the easternmost outpost of German power for over 400 years. After World War II it was called Kaliningrad and became the largest naval base of the USSR. Today Königsberg is located between Lithuania and Poland. How do the seven bridges of Königsberg look today? Do people still attempt to find the impossible path? First, let's recap the Königsberg bridge problem—

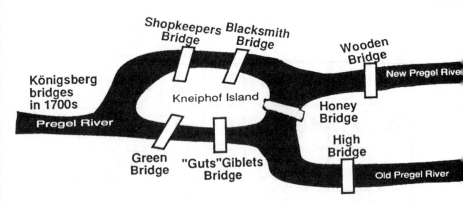

Over the centuries, the Königsberg bridge problem has provided a wealth of enjoyment and mathematical interest. The problem dates back to the 1700s. The setting was the city of Königsberg situated along the Pregel River. The two islands in the river were part of the city and connected to it by seven bridges. A delightful tradition had developed among the residents of Königsberg of taking a Sunday stroll along the city banks and islands while attempting to discover a path that would transverse all seven bridges without recrossing any bridge. Although it was considered an amusing entertainment

by most people of the time, a mathematician discovered and developed another twist to this recreational pastime. Swiss mathematician Leonhard Euler (1707-1783) had learned of the Königsberg bridge problem in St. Petersburg while in the service of Catherine the Great of Russia.

In 1735 Euler presented a paper to the Russian Academy that was more far reaching and had a more substantial impact on mathematics than simply the solution to the bridge problem. He set into motion ideas that launched the field of topology. Unlike Euclidean geometry, which deals with size, shape and rigid objects, topology is a geometry that studies properties of objects that remain unchanged even when the size and shape of the objects are distorted. For example, if a triangle is distorted into a square or a circle, topology studies which of the object's properties remain unchanged. With the Königsberg bridge problem Euler transformed and simplified a physical setting into a mathematical design (called graph or network) which encompassed and sim-plified the problem. For each part of the city to which the bridges led he assigned a vertex point and illustrat-ed each bridge with an arc. He con-cluded that the problem of crossing all seven bridges without doubling back was comparable to tracing over the network with a pencil without ever lifting it up. Euler identified each vertex as either an odd or even

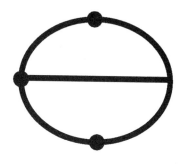

point. He noted that an even point was created by either passing in and out of the vertex **or** by beginning and ending your journey at that point. An odd vertex, on the other hand, was created by being either the starting **or** ending point of the journey. Thus any graph that was traceable (no doubling back) could only have at most 2 odd vertices— i.e. 0 if they were all even **or** 2 if one was a beginning point and one an ending point. In addition, he figured

One of the three original remaining bridges of Königsberg.

out that if the graph had an even number of odd vertices, say 10, one would have to lift the pencil half this number, that is 5 times, to trace the entire graph. In his dissertation, Euler points out that the Königsberg bridge problem seemed geometric in nature, but that Euclidean geometry did not seem to apply since the bridge problem did not deal with "magnitudes" nor could it be resolved using "quantitative calculations". Instead the problem belonged to the "geometry of position", as topology was first described and named by Gottfried Wilhelm von Leibniz. Thus, Euler's solution to the Könisgberg bridge problem acted as a catalyst and introduction to the field of topology.

As the illustration indicates, each of the seven bridges had a specific name probably related to what was on that side of the bridge. Today only three of the seven original bridges remain — the Honey bridge, the High bridge, and the Wooden bridge. A new bridgeway was constructed that connects the two banks as shown and bypasses the island of Kneiphof completely. Tour guides often relate the story of the Königsberg bridge problem. Some guides even claim it remains unsolved. If a network of the updated Königsberg bridges is drawn, the new problem loses its appeal. If the new bridge

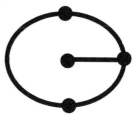

touched down on the island, the network would be somewhat more interesting. Unfortunately, the seven bridges of Königsberg are history, but the legacy this problem left is not destroyed as easily as its bridges. Euler's brilliant solution remains an important part in the development of topology.

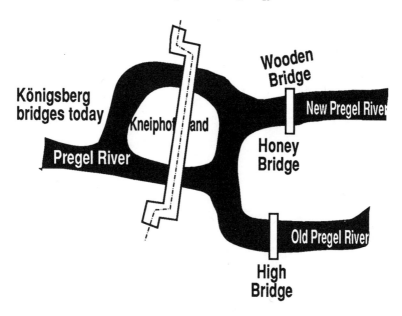

Acknowledgements: Special thanks for updated information and photograph to Art Cooley and to Rod Crittenden for bringing the information to my attention.

CHECKERBOARD MANIA

Thousands of years ago ancient Egyptians played games on checkered boards. However, the modern game of checkers dates back to the beginning of the 12th century in Europe. It used the same board as chess, playing pieces from backgammon, and followed along the line of the game *alquerque* with regard to the movement and number of pieces. The board itself poses a wealth of problems that have been developed over the years.

COVERING CHECKERBOARD WITH DOMINOES

If two opposite corners of a checkerboard are removed, can the checkerboard be covered by dominoes? Assume that each domino is the size of two adjacent squares of the checkerboard. The dominoes cannot be placed on top of each other and must lie flat.

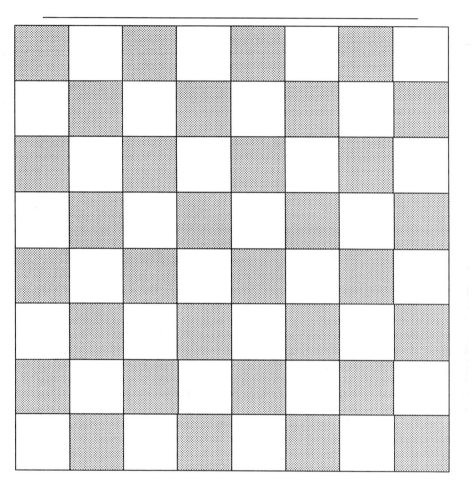

TAKING A CHECKERBOARD APART
WITH ONE FELL SWOOP

Using your visualization skills and folding techniques, determine a way to make one straight cut to separate the checkerboard into 2x2 squares, such as this one.

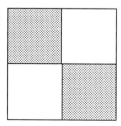

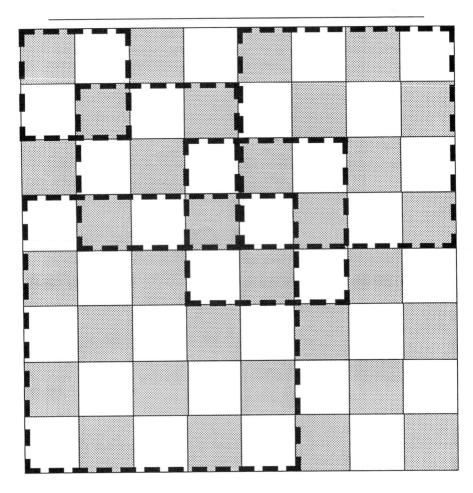

THE SQUARES OF A CHECKERBOARD

The 8 rows and 8 columns of black and red squares composing a checkerboard can be grouped into different sized squares. These squares range in size from 8x8 to 1x1. How many squares of various sizes can be found on a checkerboard?

TWO OF SAM LOYD'S CHECKERBOARD PUZZLES

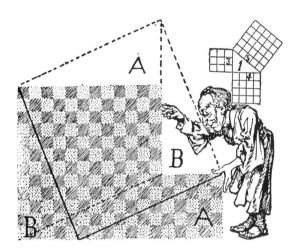

Puzzlist Sam Loyd called this puzzle *Mrs. Pythagoras' Puzzle*. You must devise a method to cut this checkered pattern into three pieces which can be joined into a 13x13 checkered square in which the pattern and direction of the design is retained.

This Sam Loyd checkerboard puzzle was called *A Battle Royal*. Rearrange the 8 pieces so that they form an 8x8 checker board.

A FEW OLDIES

SAM LOYD'S GET OFF THE EARTH PUZZLE

Sam Loyd's *Get Off the Earth* puzzle is one of the most popular disappearing puzzles ever made. Loyd patented it in 1896, and sold more than 10 million copies.

The puzzle was made with a movable circular disk that rotated the Earth. Can you discover what happens to the 13th warrior when the sphere is rotated from NE to NW?

SAM LOYD'S BATTLE OF THE FOUR OAKS PUZZLE

Four trees are in a line on a square field. Divide the field into four pieces so that each piece is the same size and shape and has one oak tree.

SAM LOYD'S WEIGHT OF A BRICK PUZZLE

If a brick balances with three quarters of a brick and three quarters of a pound, then how much does the brick weight?

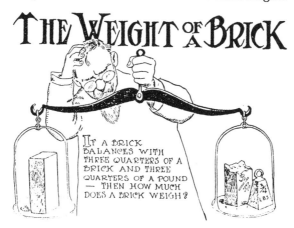

SAM LOYD'S CHEESE PROBLEM

If the cheese round were cut as shown, how many wedges of cheese would be produced?

WHAT'S LEFT IN THE CANDY BOXES ?

Somehow the labels of these boxes of candies got mixed up so that no label now indicates the correct contents.

What is the least number of candies you must test from which box(es) to determine which box has what?

3 chocolates

3 cremes

2 chocolates 1 creme

LEWIS CARROLL'S THE TWO CLOCK PUZZLE

Which is better, a clock that is right only once a year or a clock that is right twice a day?

TARTAGLIA'S PUZZLE

16th century Italian mathematician Niccolò Tartaglia is famous for discovering the solution to the general cubic equation (1535). He also wrote the three volume set *Trattato generale di numeri et misure (Treatise on Numbers and Measures* (1556-1560). In addition, he was the first to translate Euclid's *The Elements* into a modern Western language (1543). Here is a very famous mathematical teaser he devised:

Three newlyweds come to a river bank where a small boat is to take them across, but the boat can hold only two people at a time.   Each husband is the jealous type and very protective of his beautiful bride.   In order to keep things mellow, they decide that no woman is to be left with a man unless her husband is present.

How do the couples get to the other side of the river? What's the minimum number of trips?

page 31. **WHY ARE MANHOLES ROUND?** SOLUTION: A circle by definition is the set of all points in a plane that are equidistant from a given point. Therefore, a circular lid could not fall through a circular manhole, since all its diameters are equal. If the manhole were square, its lid could fall into the manhole by putting it through along the square's diagonal.

page 171. **EULER'S MAGIC FORMULA**
SOLUTION: For a rhombicosidodecahedran— F=62 (30 squares, 20 triangle, and 12 pentagons); V=60; E=120; therefore F+V-E=2.

page 271. **WHAT ARE THE SCENARIOS?**
SOLUTION: (1) Tom and Jerry are professional baseball players, but play on different teams. This particular day their teams were playing each other. Tom is a catcher who prevented Jerry from reaching home plate.
(2) While walking down the street, Eric had a terrible attack of hiccups. He went into the bar to get a glass of water to see if it would help him. The bartender, realizing Eric's problem, thought he would try to scare the hiccups out of Eric. It worked!
(3) Mary's husband had died a number of years ago. She kept his ashes in an urn on her kitchen table. Also on her table was a bowl with goldfish. Mary had left her kitchen window open this particular day. Her neighbor's cat had come in, and in trying to get the goldfish, the cat knocked over the goldfish bowl and the urn. Her husband's ashes were spilled onto the floor.

page 272. **RECTANGULAR VOLUME ILLUSION**
SOLUTION: There are sixty-one 2x1x1 blocks in the structure.

page 273. **THE IMPOSSIBLE POSSIBLE**
SOLUTION: The slit that is cut in the paper is placed into the hole. Then the buttons can be fed easily through the slit and removed from the paper.

page 273. **STACKING THE DIE**
SOLUTION: The trick to this problem is recalling that opposite faces of a die total 7. The first stack has 10 dice which would give 70 pips minus the face of the top die that is visible, gives 69 pips. The second stack has 7 dice, giving 49 pips, minus the top die's pips of 3 gives 46. The grand total is therefore 69+46=105 pips.

page 275. **THE BLOCK LETTER PUZZLE**
SOLUTION:

page 276. **THE MILKMAN'S PUZZLE**
SOLUTION:

10qt.	10qt.	5qt.	4qt.
10	10	0	0 – starting
5	10	5	0
5	10	1	4
9	10	1	0
9	10	0	1
4	10	5	1
4	10	2	4
8	10	2	0
8	6	2	4
10	**6**	**2**	**2** – ending

page 277. IS IT POSSIBLE?
SOLUTION:

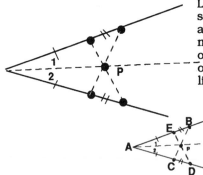

Locate the 4 points on the sides of the angle as shown, and drawn in the two segments. The segments' point of intersection and the vertex of the angle determine the line that bisects the angle.

Note: Show∠1≅∠2 by first showing
ΔABC≅ΔADE by SAS
→∠B≅∠D→ΔBEP≅ΔDCP
by SAA→BP≅DP
→ΔBPA≅ΔDPA by
SSS→∠1≅∠2

page 277. CAN YOU MAKE A LATIN SQUARE?
SOLUTION:

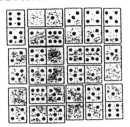

page 285. PLAYING WITH POLYHEXES
SOLUTION: The triangle is the only figure that cannot be polyhexed.

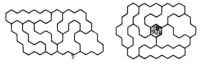

page 286. LEWIS CARROLL'S DOUBLETS
SOLUTION: (1) EYE—LYE—LIE—LID
(2) PITY—PITS—FITS—FINS—FIND—FOND—FOOD—GOOD
(3) TREE—FREE—FLEE—FLED—FEED—WEED—WELD—WOLD—WOOD
(4) OAT—RAT—ROT—ROE—RYE

page 289. CURRY TRIANGLE EXPLANATION
SOLUTION: When the other missing parts are calculated using similar triangles, the results would actually create sections that would overlap in some places and leave gaps in others.

page 290. THE SQUARE TRANSFORMATION
SOLUTION:

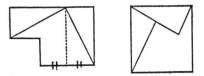

page 302. COVERING THE CHECKERBOARD WITH DOMINOES
SOLUTION: It is not possible to cover the altered checkerboard with dominoes. A domino must occupy a red and black square. Since both corners removed were the same color, there will not be a compatible number of red and black squares left.

page 303. **TAKING A CHECKERBOARD APART**
SOLUTION:

Step 1   Step 2

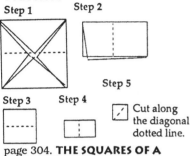

Step 5

Step 3   Step 4

Cut along the diagonal dotted line.

page 304. **THE SQUARES OF A CHECKERBOARD**
SOLUTION: There are:
1 square of size 8x8
4 square of size 7x7
9 squares of size 6x6
16 squares of size 5x5
25 squares of size 4x4
36 squares of size 3x3
49 squares of size 2x2
64 squares of size 1x1
for a total of 204 squares

page 305. **A BATTLE ROYAL**
SOLUTION:

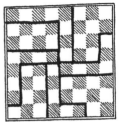

page 307. **THE FOUR OAKS PUZZLE**
SOLUTION:

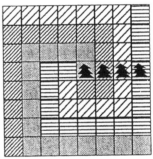

page 305. **MRS. PYTHAGORAS' PUZZLE**
SOLUTION:

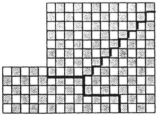

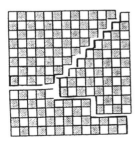

page 307. **THE WEIGHT OF A BRICK PUZZLE**

SOLUTION: One way to approach the problem is to add three more bricks to the left side of the scale. To balance this quadruple the items on the right side. This means 4 bricks on the left balance a total of 3 bricks and 3 pounds. This produces the equation 4B=3B+3.

Solving for B (brick's wight in pounds), we get B=3.

Another approach is to write an equation from the first scale's balance,

namely B=(3/4)B+(3/4)

and solve for B.

page 308. **THE CHEESE PROBLEM**

SOLUTION: According to Sam Loyd, the first cut divides the cheese into two pieces. The second cut makes 4 pieces. The third makes 8; the fourth 15; the fifth 26; and the sixth 42.

page 308. **WHAT'S LEFT IN THE CANDY BOXES?**

SOLUTION: You only have to test one candy from the box labeled *2 chocolates & 1 cream*. If the candy you test is chocolate, you know it cannot have 2 chocolates & 1 cream because the box is mislabeled. Therefore, it must have 3 chocolates. Since all boxes are mislabeled, this would mean the box labeled *3 creams* must have 2 chocolates & 1 cream and the box labeled *3 chocolates* has 3 creams. If the candy you test is a cream, you know that the box has 3 creams. This would mean the box labeled *3 creams* must have 3 chocolates, and the box labeled *3 chocolates* has 2 chocolates & 1 cream.

page 308. **LEWIS CARROLL'S CLOCK PUZZLE**

SOLUTION: *Which is better, a clock that is right only once, or a clock that is right twice everyday? "The latter," you reply, "unquestionably." Very*

good, now attend.

I have two clocks: one doesn't go at all, and the other loses a minute every day; which would you prefer? "The losing one," you answer, "without a doubt." Now observe: the one which loses a minute a day has to lose twelve hours or seven hundred and twenty minutes, before it is right again, consequently it is only right once in two years, whereas the other is evidently right as often as the time it points to comes around, which happens to be twice a day. — Lewis Carroll

page 308. **TARTAGLIA'S PUZZLE**

SOLUTION: The rule that no woman is to be left with a man unless her husband is present makes 11 crossings necessary.

Notation: husband-one=H1; wife-one-W1 etc.

1) H1 takes W1 over and returns— 2 crossings

2) W3 takes W2 over and returns— 2 crossings

3) H2 takes H1 over and returns with W2— 2 crossings

4) H3 takes H2 over and returns— 2 crossings

5) H3 takes W3 over and H2 returns to pick up his wife.— 3 crossings.

Andrews, W.S.. *Magic Squares and Cubes*, Dover Publications, Inc., New York, 1960.

Ball, W.W. Rouse and Coxeter, H.S.M.. *Mathematical Recreations and Essays*, Dover Publications, Inc., New York, 1987.

Ball, W.W.Rouse. *A Short Account of the History of Mathematics*, Dover Publications, Inc., New York, 1960.

Banchoff, Thomas F. *Beyond The Third Dimension*, W.H. Freeman & Co., New York, 1990.

Barnsley, Micahel. *Fractals Everywhere*, Academic Press, Inc., Boston, 1988.

Barrow, John D.. *Pi In The Sky*, Oxford University Press, Oxford, 1992.

Bell, Eric Temple. *The Magic Of Numbers*, Dover Publications, Inc., New York, 1946.

Bell, R.C.. *Board and Tables Games from Many Civilizations*, Dover Publications, Inc., New York, 1979.

Bell, R.C.. *Discovering Old Board Games*, Shire Publications Ltd., United Kingdom, 1973.

Blackwell, William. *Geometry & Architecture*, Key Curriculum Press, Berkeley, CA, 1994.

Bold, Benjamin. *Famous Problems of Geometry*, Dover Publications, Inc., New York, 1969.

Boles, Martha and Newman, Rochelle. *Universal Patterns*, Pythagorean Press, Bradford, MA 1987.

Bool, F.H.; Krist, J.R.; Locher, J.L.; Wierda, F.. *M.C. Escher His Life & Complete Works*, Harry N. Abrams, New York, 1982.

Boyer, Carl. *A History of Mathematics*, Princeton University Press, New Jersey, 1985.

Briggs, John and Peat, David. *Turbulent Mirror*, Harper & Row, New York, 1989.

Brizio, A.M.; Brugnoli, M.V.; Chastel, A.. *Leonardo the Artist*, McGraw-Hill Co., New York, 1980.

Bunch, Bryan H.. *Mathematical Fallacies and Paradoxes*, Van Nostrand Reinhold Co., New York, 1982.

Campbell,D.M. and Higgins, J. editors. *Mathematics: People, Problems, Results*, Wadsworth International, Belmont, CA 1984.

Clark, Don. *Myers Laboratories; Hearing in Three Dimensions*, San Francisco Chronicle, San Francisco, June 11, 1987.

Cocteau. Jean. *Flight*, Hill & Wang, New York, 1963

Cook, Theodore A.. *The Curves of Life*, Dover Publications, Inc., New York, 1979

Coxeter, H.S.M.; Emmer, M.; Penrose, R.; Teuber,M.L., editors. *M.C. Escher: Art & Science*, North-Holland, Amsterdam, 1986.

Crandall, Richard E. *Projects in Scientific Computation*, Springer-Verlag Publishers, Santa Clara, CA, 1994.

Davies, W.V.. *Reading the Past — Egyptian Hieroglyphs*, University of California Press, Berkeley, CA, 1987.

Davis, Philip and Hersh, Reuben. *The Mathematical Experience*, Houghton Mifflin Co., Boston, 1981.

Davis, Philip J. and Hersh, Reuben. *Descartes' Dream*, Houghton Mifflin Co., Boston, 1986.

Davis, Philip J. and Hersh, Reuben. *The Mathematical Experience*, Houghton Mifflin Co., Boston, 1981.

Dewdney, A.K.. *The Magic Machine*, W.H.Freeman & Co., New York, 1990.

Dilke, O.A.W.. *Reading the Past—Mathematics & Measurment*, University of California Press, Berkeley, CA, 1987.

Dunham, William. *Journey Through Genius*, John Wiley & Sons, New York, 1990.

Edwards, Edward B.. *Patterns and Design with Dynamic Symmetry*, Dover Publications, Inc., New York, 1967.

Ernst, Bruno. *The Magic Mirror of M.C. Escher*, Ballantine Books, New York, 1976.

Ferguson, Claire. *Helaman Ferguson: Mathematics in Stone and Bronze*, Meridian Creative Group, Erie, PA, 1994.

Ferrell, J.E.. *Claude Phené—Microcomputers & Micro-irrigation*, Computerland Magazine, July/August, 1990.

Forsyth, Andrian. *The Architecture of Animals*, Camden House, Ontario, 1989.

Freias, Bill. *The Urban Forest*, Computerland Magazine, Jan./Feb. 1989.

Frisch, Karl von. *Animal Architecture*, Van Nostrand Reinhold Co., New York,1974.

Gamow, George. *One, Two, Three ...Infinity*, The Viking Press, New York, 1947.

Garcia, Linda.*The Fractal Explorer*, Dynamic Press,Santa Cruz,CA 1991.

Gardner, Martin. *Codes, Ciphers and Secret Writing*, Dover Publications, Inc., New York, 1972.

Gardner, Martin. *Fractal Music, Hypercards and more...*, W.H.Freeman & Co., New York, 1991.

Gardner, MArtin. *Knotted Doughnuts*, W.H. Freeman & co., New York, 1986.

Gardner, Martin. *Mathematics, Magic & Mystery*, Dover Publications, Inc., New York, 1956.

Gardner, Martin. *Time Travel*, W.H. Freeman & Co., New York, 1987.

Gardunkel, Solomon & Steen, Lynn A. editors. *For All Practical Purposes*, W.H.Freeman & Co., New York, 1991.

Ghyka, Matila. *The Geometry of Art and Life*, Dover Publications, Inc., New York, 1977.

Gleason, Norma. *Cryptograms & Spygrams*, Dover Publications, Inc., New York, 19881.

Gleick, James. *Chaos*, Penguin Group, New York, 1987.

Glenn, William H. and Johnson, Donovan A.. *Invitation to Mathematics*, Doubleday & Co., Inc., Garden City, NY, 1961.

Golos, Ellery B.. *Foundations of Euclidean and Non-Euclidean Geometries*, Holt, Rinehart & Winston, Inc., New York, 1968.

Greensberg, Marvin Jay. *Euclidean and Non-Euclidean Geometries*, W.H. Freeman & Co., New York, 1980.

Grübaum, Branko and Shephard, G.C.. *Tillings and Patterns*, W.H.Freeman and Co., New York, 1987.

Grunfeld, Frederic V. ed.. *Games of the World*, Holt, Rinehart & Winston, Inc., New York, 1975.

Hall, Rupert A.. *From Galileo To Newton*, Dover Publications, Inc., New York, 1981.

Hall, Stephen S. *A Molecular Code Links Emotions*, Smithsonian Magazine, Smithsonian Asso., Washington D.C., 1990.

Hambridge, Jay. *The Elements of Dynamic Symmetry*, Dover Publications, Inc., New York, 1967.

Hawkins, Gerald s.. *Mindsteps to the Cosmos*, Harper & Row, Publishers, NY, 1983.

Hemmings, Ray and Tanta, Dick. *Images of Infinity*, Leapfrogs, United Kingdom, 1984.

Herrick, Richard, editor. *The Lewis Carroll Book*, Tudor Publishing Co., New York, 1944.

Heydenreich, L.H.; Dibner, B.; Reti, L.. *Leonardo the Inventor*, McGraw-Hill Co., New York, 1980.

Hoffman, Paul. *Archimedes' Revenge*, W.W.Noron & Co., Inc., New York, 1988.

Hollindale, Stuart. *Makers of Mathematics*, Penquin Books, London, 1989.

Ifrah, Georges. *From One to Zero*, Penquin Books, New York, 1985.

Italo Cavino. *Cosmi-comics*, Harcourt Brace Jovanovich, New York, 1965.

Ivins, Jr., William M.. *Art & Geometry*, Dover Publications, Inc., New York, 1946.

Johnson, Donovan A. and Glenn, William H.. *Exploring Mathematics On Your Own*, Doubleday & Co., Inc., New York, 1961.

Jones, Richard Foster. *Ancient and Moderns*, Dover Publications, Inc., New York, 1981.

Kappraff, Jay. *Connections*, McGraw-Hill Inc., New York, 1991.

Kasner, Edward and Newman, James R.. *Mathematics and the Imagination*, Microsoft Press, Redmond, WA, 1989.

Kline, Morris. *Mathematical Thought From Ancient to Modern Times*, Oxford University Press, New York, 1972.

Kline, Morris. *Mathematics, The Loss of Certainty*, Oxford University Press, New York, 1980.

Kosko, Bart. *Fuzzy Thinking*, Hyperion, New York, 1993.

Kraitchik, Maurice. *Mathematical Recreation*, Dover Publications, Inc., New York, 1953.

Lebow, Irwin. *The Digital Connection*, Computer Science Press, New York, 1991.

Lloyd, Steon; Rice, David; et al. *World Architecture*, McGraw-Hill Co., London, 1978.

Loon, Broin van. *DNA—The Marvelous Molecule*, Tarquin Publications, Stradbroke, England, 1990.

Loyd, Sam. *Cyclopedia of Puzzles*, Sam Loyd, 1914.

Luckiesh, M.. *Visual Illusions*, Dover Publications, Inc., New York, 1965.

Lysing. *Secret Writing*, Dover Publications, Inc., New York, 1974.

Mandelbrot, Benoit. *The Fractal Geometry of Nature*, W.H.Freeman & Co., New York, 1983.

Maucaulay, David. *The Way Things Work*, Houghton Mifflin, Boston, 1988.

Menninger, K.W.. *Mathematics In Your World*, The Viking Press, New York, 1954.

Moran, Jim. *The Wonders of Magic Squares*, Random House, New York, 1982.

Neugebauer, O.. *The Exact Science In Antiquity*, Dover Publications Inc., New York, 1969.

Newman, James R., et. al.. *The World of Mathematics*, Simon & Schuster, New York, 1956.

Nicolis, Grégoire and Prigogine, Ilya. *Exploring Complexity*, W.H. Freeman and Co., New York, 1989.

Nobles, Barr. *Latest Trick in Recording — QSound*, San Francisco Chronicle, San Francisco, Feb. 6, 1991.

Ogilvy, C. Stanley and Anderson, John T.. *Excursions in Number Theory*, Dover Publications, Inc., New York, 1966.

Pagels, Heinz R.. *Perfect Symmetry*, Simon & Schuster, New York, 1985.

Palfrreman, Jon and Swade, Doron. *The Dream Machine*, BBC Books, London, 1991.

Pappas, Theoni. *Mathematics Appreciation*, Wide World Publishing/Tetra, San Carlos, CA, 1987.

Pappas, Theoni. *Math Talk,* Wide World Publishing/Tetra, San Carlos, CA, 1991.

Pappas, Theoni. *More Joy of Mathematics,* Wide World Publsihing/Tetra, San Carlos, CA, 1993.

Pappas, Theoni. *The Joy of Mathematics,* Wide World Publishing/Tetra, San Carlos, CA, 1993.

Pappas, Theoni. *What Do You See?— An optical illusion slide show,* Wide World Publishing/Tetra, San Carlos, CA, 1989.

Paulos, John Allen. *Beyond Numeracy,* Alfred A. Knopf, New York, 1991.

Pedoe, Dan. *Geometry and the Visual Arts,* Dover Publications, Inc., New York, 1976.

Penrose, Roger. *The Emperor's New Mind,* Oxford University Press, New York, 1989.

Peterson, Ivars. *Looking-Glass Worlds,* Science News, Washington D.C., Jan. 4, 1992.

Peterson, Ivars. *The Mathematical Tourist,* W.H. Freeman & Co., New York, 1988.

Peterson, Ivars. *The Mathematical Tourist,* W.H. Freeman and Co., New York, 1988.

Peterson, Roger T. & editors of Life. *The Birds,* Time, Inc., New York, 1963

Pevsner, Nicholaus; Felming, John; Honour, Hugh. *A Dictionary of Rachitecture,* The Overlook Press, Woodstock, New York 1976.

Pickover, Clifford A.. *Computers, Patterns, Chaos, & Beauty,* St. Martin's Press, New York, 1990.

Pickover, Clifford. *Computers and the Imagination,* St. Martin's Press, New York, 1991.

Pickover, Clifford. *Mazes of the Mind,* St. Martin's Press, New York, 1992.

Pierce, John R.. *The Science of Musical Sound,* W.H.Freeman & Co., New York, 1983.

Prigogine, Ilya. *Order Out Of Chaos,* Bantam Books, Toronto, 1988.

Ransom, William R.. *3 Famous Geometries,* William Ransom, 1959.

Ransom, William R.. *Can and Can't of Geometry,* J. Weston Walch, Portland, ME, 1960.

Resnikoff, H.L. and Wells, Jr., R. o.. *Mathematics in Civilization,* Dover Publications, New York, 1973.

Rheingold, Howard. *Virtual Reality,* Summit Books, New York, 1991.

Richter, Jean Paul. *The Notebooks of Leonardo da Vinci,* Dover Publications, New York, 1970.

Robbin, Tony. *Computers, Art and the 4th Dimension,* Little Brown & Co., Boston, 1992.

Robins, Gay and Shute, Charles. *The Rhind Mathematical Papyrus*, Dover Publications, Inc., New York, 1987.

Rosenberg, Nancy. *How To Enjoy Mathematics with Your Child*, Stein and Day, New York, 1970.

Rosenberg, Scott. *Virtual Reality Check*, San Francisco Examiner, San Francisco, April 19, 1990.

Rucker, Rudy. *Infinity & the Mind*, Bantam Books, New York, 1982.

Ruelle, David. *Chance and Chaos*, Princeton Uninersity Press, New Jersey, 1991.

Schattschneider, Doris. *Visions of Symmetry*, W.H. Freeman, New York, 1990.

Schimmel, Annemarie. *The Mystery of Numbers*, Oxford University Press, Oxford, 1992.

Schroeder, Manfred. *Fractals, Chaos, Power Laws*, W.H.Freeman and Co., New York, 1990.

Sharpe, Richard and Piggott, John, editors. *The Book of Games*, Galahad Books, New York, 1977.

Shlain, Leonard. *Art & Physics*, William Morrow and Co., NY, 1991.

Smith, D.E.. *History of Mathematics*, Dover Publications, Inc., New York, 1951.

Smith, Laurence Dwight. *Cryptography*, Dover Publications, Inc., New York, 1943.

Steen, Lynn Arthur, ed., *Mathematics Today*, Vintage Books, NY, 1980.

Stevens, Peter S. *Patterns In Nature*, Little Brown & Co., Boston, 1974.

Stewart, Ian and Golubitsky, Martin. *Fearful Symmetry-Is God a Geometer?* Blackwell Publishers, Oxford, 1992.

Stewart, Ian. *Another Fine Math You've Got Me Into....* W.H. Freeman & Co., New York, 1992.

Stewart, Ian. *Does God Play Dice?* Basil Blackwell, Oxford, 1989.

Struik, Dirk. *A Concise History of Mathematics*, Dover Publications Inc., New York, 1967.

Sutton, O.G.. *Mathematics in Action*, Dover Publications, NY, 1957.

Trachtenberg, Marvin and Hyman, Isabelle. *Architecture from Prehistoric to Post Modernism*, Harry N. Abrams, Inc., NY, 1986.

Tuller, Annita. *A Modern Introduction to Geometries*, D. Van Nostrand Co. Inc., Princeton, New Jersy, 1967.

Waerden, B.L. van der. *Science Awakening*, John Wiley & Sons, Inc., New York, 1963.

Waldrop, M. Mitchell. *Complexity*, Simon & Schuster, New York, 1992.

Wolf, Fred Alan. *Parallel Universes*, Simon & Schuster, NY, 1988.

Zammatio,c.; Marinoni,A.; Brizio,A.M.. *Leonardo the Scientist*, McGraw-Hill Co., New York, 1980.

Mathematics teacher and consultant Theoni Pappas received her B.A. from the University of California at Berkeley in 1966 and her M.A. from Stanford University in 1967. Pappas is committed to demystifying mathematics and to helping eliminate the elitism and fear often associated with it.

In addition to *The Magic of Mathematics*, her other innovative creations include *The Math-T-Shirts*, *The Mathematics Calendar*, *The Children's Mathematics Calendar*, *The Mathematics Engagement Calendar*, and *What Do You See?* — an optical illusion slide show with text. Pappas is also the author of the following books: *The Joy of Mathematics*, *More Joy of Mathematics*, *Mathematics Appreciation*, *Math Talk—Mathematical Ideas in Poems for Two Voices*, and *Fractals, Googols & other Mathematical Tales*.

Mathematics Titles
by Theoni Pappas

THE MAGIC OF MATHEMATICS
$10.95 • 224 pages•available Spring'94
illustrated•ISBN:0-933174-99-3

FRACTALS, GOOGOL, and Other
Mathematical Tales?
$9.95 • 64 pages • for all ages
illustrated•ISBN:0-933174-89-6

THE JOY OF MATHEMATICS
$10.95 • 256 pages
illustrated•ISBN:0-933174-65-9

MORE JOY OF MATHEMATICS
$10.95 • 306 pages
cross indexed with *The Joy of Mathematics*
illustrated•ISBN:0-933174-73-X

MATHEMATICS APPRECIATION
$10.95 • 156 pages
illustrated•ISBN:0-933174-28-4

MATH TALK
mathematical ideas in poems for two voices
$8.95 • 72 pages
illustrated•ISBN:0-933174-74-8

THE MATHEMATICS CALENDAR
$8.95 • 32 pages • written annually
illustrated

THE CHILDREN'S MATHEMATICS CALENDAR
$8.95 • 32pages • written annually
illustrated

WHAT DO YOU SEE?
An Optical Illusion Slide Show with Text
$29.95 • 40 slides • 32 pages
illustrated•ISBN:0-933174-78-0